AN INTRODUCTION TO MOORE-PENROSE RINGS

AN INTRODUCTION TO MOORE-PENROSE RINGS

Volume I

Gregory Battle, Ph.D.

iUniverse, Inc.
New York Lincoln Shanghai

AN INTRODUCTION TO MOORE-PENROSE RINGS
Volume I

iUniverse books may be ordered through booksellers or by contacting:

iUniverse
2021 Pine Lake Road, Suite 100
Lincoln, NE 68512
www.iuniverse.com
1-800-Authors (1-800-288-4677)

ISBN-13: 978-0-595-37806-7 (pbk)
ISBN-13: 978-0-595-82186-0 (ebk)
ISBN-10: 0-595-37806-4 (pbk)
ISBN-10: 0-595-82186-3 (ebk)

Printed in the United States of America

Contents

PREFACE

This Volume I book introduces the theoretical structure of Moore-Penrose rings (named in honor of the mathematicians Edwin H. Moore & Roger Penrose) prevalent in the algebraic literature in various disguises or representations. This book represents a tour de force for introducing so many topics and elementary ring theory concepts embedded in the exciting variety of Moore-Penrose rings. Written in a clear and explanatory style, this algebraic book is recommended as a supplement for an undergraduate course as well as for engaging independent study on the graduate level. For example, the text is well written to illustrate with logical ease how the Moore-Penrose ring structure of a ground ring can be extended to an ambient matrix ring. This self-contained and rigorous introduction to Moore-Penrose rings is the unique algebra book that gives undergraduate algebra students examples of how ring theory concepts (such as generating ideals and constructing integral extensions) can be applied on both the theoretical level as well as in the real world with specific scientific applications where Moore-Penrose rings are in plentiful usage. Many of the ideas contained in this book were borne in outstanding research conducted on Moore-Penrose rings for eighteen years at Washington University in St. Louis under a mentor who produced impressive ring theory papers coached by the renown algebraist, I. N. Herstein. Also, the author has worked with atmospheric scientists in numerical weather prediction at the National Center for Atmospheric Research in Boulder, Colorado, and at the Air Force Research Laboratory on Hanscom Air Force Base where many applications using the Moore-Penrose matrix equations proved to be beneficial in solving complex problems simulating meteorological phenomena. There are some great classic books on either the generalized inverse matrix, or the pseudoinverse matrix X which satisfies the Moore-Penrose equations

$$1.\ AXA = A \qquad 2.\ XAX = X \qquad 3.\ (XA)^* = XA \qquad 4.\ (AX)^* = AX$$

Where the matrix A represents some linear operator on a real or complex linear space, and the involution * is the standard complex-conjugate transpose rule for matrices. A couple of these books are (a) Generalize <u>Inverses: Theory and Applications</u>, A. Ben-Isreal and T. N. E. Greville; New York, John Wiley & Sons, 1977 and (b) <u>Generalized Inverse of Matrices and Its Applications</u>, C.R. Rao and S. K. Mitra; New York, Johan Wiley & Sons, 1971. Both these books contain some theoretical development of the algebraic structure that encompass the matrix algorithms rendered for obtaining a generalized inverse matrix under various restrictions of rank, conditioning, symmetry, and accessible computer processing memory. This "Introduction to Moore-Penrose Rings" details their ambient structural framework as a canvas from which all the aforementioned mathematicians have produced some significant numerical and computational mathematics literature. However, this introductory theoretical treatise on thef characterizations of Moore-Penrose rings does not require constant referring to the empirical constructions of other great thinkers to justify its desirability as a comprehensive, yet elegant analysis of the texture of a genre of rings that should remain prominent for future generations of applied mathematics technicians. Volume Two of this algebraic book series will be mostly pedantic proofs of some startling structure characterizations of Moore-Penrose Rings.

To illustrate the diverse applications of the theory of Moore Penrose rings, the following examples are offered as humble testimony to beckon more scrutiny of algebraic rings that have the fruition to be harvested with continual and startling yield:

<u>Example 1</u>: The meteorologist Paul R. Julian and linear algebraist Alan Cline, used the method of least squares to estimate the spectral ordinates for generating isothermal curves on equally and unequally space grids in their research in the Numerical Weather Prediction group back in the early 1970's at the National Center for Atmospheric Research. The author worked under their tutelage as an undergraduate intern. The operative equation was $A\vec{p} = \vec{r}$ where A was the matrix of cosines from a Fourier cosine series, $\vec{p}$ was the vector of spectral ordinates to be determined, and $\vec{r}$ were the temperature data from a weather observation balloon taken at irregular mapping coordinates. The computational solution for an ill-conditioned matrix A was $\vec{p} = A^+ \vec{r}$ where A^+ represented the pseudoinverse of the matrix A.

<u>Example 2</u>: The Moore-Penrose generalized inverse may be used to find an approximate inverse of the Jacobian of the collision term J in the linearized finite difference quantum Boltzman equation

$$f(x \pm \delta x, t + \delta t - f(x, t)) = J\delta f(x,t)$$

where the term J contains particles of the quantum equilibrium function and f is the vector of occupation probabilities. It turns out that the Moore-Penrose generalize inverse of J is $J^g = (\dfrac{1}{J_+ - J_-})$ J where J_+, J_- are partials of the equi-librium condition function Ω. This example is taken from the yet unpublished article <u>Physical Review</u>, "Open Quantum System Model Of The One-Dimensional Burger Equation With Arbitrary Sheer Viscosity", Jeffrey Yepez, submitted 2005 from the Air Force Research Laboratory at Hansom Air Force Base near Waltham, Massachusetts.

Of course there are many more examples of least square methods to solve both engineering and large-scale matrix economic models. Volume III of this *Introduction To Moore-Penrose Rings* shall contain applications and computa-tional techniques for the applied mathematician or physicist to relish.

To get a complete characterization of Moore-Penrose rings, the author gener-alized the matrix Moore-Penrose equations above into the following relations for an arbitrary nonzero ring element x

1. $axa - a$ 2. $xax = x$ 3. $(xa)^* = xa$ 4. $(ax)^* = ax$

where a is an arbitrary nonzero ring element and the operation * represents a ring involution. (of which matrix transpose is a particular case). All the rings investigated in this book are approached as a class according to which one of the four properties above are satisfied. All rings that satisfy the first property will be called MP1; all rings that satisfy the second property will be called MP2, etc. Not much theoretical development was necessary for the MP1 rings as they are widely known in the algebraic world as Von Neuman regular rings. Their algebraic structure has been studied extensively in the past 40 years, particularly by K. R. Goodearl whose reference is provided as a courtesy for excellent reading on the subject: <u>Von Neumann Regular Rings</u>, 2nd Edition, K.R. Goodearl, Robert E. Kreiger Publishing Co., Inc., Malabar, FL, 1991. The outline of this introduction book follows a pedantic pattern: first, MP1 or Von Neumann regular rings are discussed both theoretically and illuminated with many examples. Basic questions such as does the MP1 structure get absorbed into the quotient rings, or conveyed under canonical homomorphisms are answered either by examples or cogent proofs. Hereabout this book, the avid reader may find some ring structure investigations conducted on integral extensions. One conjecture that MP2 rings and MP1 rings are identical was answered with a cleverly-engineered counterexample—which insures that MP1 rings are only a proper subclass of the MP2 types. What good would a modern algebra book be without the use of the chain conditions on ideals as espoused by the renown mathematician Emily Nöether. Some characterizations of Nöetherian MP2 rings may be found in Chapter III of this book, and almost exclusively in Chapter IV.

The latter half of this book formulates the theoretical composition of MP3 and MP4 rings whose core elements of interest are found in a symmetric set S. Some unpublished, yet fascinating attempts, have been made by the author to classify MP3 and MP4 rings based on equivalence classes indexed in the character of the symmetric sets in such rings. Both ring types are important on the operator theory level, as they provide sufficient conditions to transform a linear operator into a symmetric mapping---which has implications on representing coordinates as wells as obtaining eigenfunctions to quantify the state of physical processes. While only a modicum of modular theory is surveyed in this incipient narrative of Moore-Penrose rings, which hopefully will be the first in a series of lengthy volumes, the end of Chapter III offers some comment on the nonsingular status of MP2 rings viewed as modules. There is a brief discussion of abelian MP2 rings which may provide insight into answering some unsolved structural problems over unit regular rings. One fundamental question from collegiate mathematics teachers and scholars may be: How does an introduction to these Moore-Penrose rings fit into the overall schema to get an algebra student very excited about learning pure mathematical concepts? The best answer proffered by this author is that the examples alone in this introduction should be sufficient to compel the inexperienced college student to explore the geologic landscape of ring theory as it offers some fantastic logic systems that clarifies groups of abstract objects that will be encountered throughout a personal education odyssey. The persistence of pristine logic throughout most mathematical subjects is clear indication that systematic reasoning creates a mental foundation upon which to engineer new rings, or to disassemble their structures in meaningful blocks that allows the reader to enjoy them with rapt visualization and categorical identification. The author encourages collegiate and graduate professors alike to use this "An Introduction to Moore-Penrose Rings" to increase the appeal of studying abstract algebra for its rich content in chiseling out essential structure and its beauty in analyzing formal logic behind broadly applied numerical algorithms.

ACKNOWLEDGMENTS

Foremost, I thank Mr. Roger Battle, Sr. and Viola Battle, my beloved parents for raising me with the great expectations of splendid accomplishment. Their hard work and tender charity I remember fondly as exemplary grace that constitutes the character of my daily strivings. My mathematical imagination soared under the guidance of Dr. Cleon Yohe at Washington University in St. Louis, Missouri. His meticulous logic examination of any algebraic problem was a model for serious theoretical development emulated to this date in my creative mathematical pursuits. Dr. Cleon Yohe's talents were developed in the Herstein group at the University of Chicago that included many engaging dialogues with the algebraists Joseph Gallian and Lance Small to mention a few. I further express deepest gratitude to my professional mentor, Mr. Cliff Daniels, and the computational genious of geodesy, Mr. Donald A. Richardson for their inspiration and endearing encouragement to push the maturation of my talents beyond unbreakable speeds. Of course, my St. Louis posse, Mr. Jeffrey Pitts, Mr. Arthur C. Jackson, Jr., and "Dr. K", Koje T. Willis are forever my friends.

I. INTRODUCTION

IA. GENERAL HISTORICAL REMARKS

Ever since the term "pseudoinverse" was coined by Max Woodbury, the notion of a generalized matrix inverse appeared in great detail with much prominence in the mathematical literature. E. H. Moore (1920), in his paper, "On The Reciprocal Of The General Algebraic Matrix"[1], defined this concept of a matrix inverse to apply to any non-square matrix of order m x n (m ≠ n). Even more, R. Penrose (1955) advanced Moore's work on a generalized inverse[2] by showing that for any matrix A over a field, there is a unique matrix X (the generalized inverse, or pseudoinverse) which satisfies the following four conditions:

$$AXA = A \ (1) \quad XAX = X \ (2) \quad (XA)^* = XA \ (3) \quad (AX)^* = AX \ (4)$$

where ( )* denotes the conjugate transpose. The matrix equations above will be called the *Moore-Penrose conditions*. There has been a proliferation of research articles and algebra textbooks devoted to computing techniques, matrix properties and statistical applications of a generalized inverse. For the moment, call a matrix A which satisfies condition (1) above MP1. A short treatise on a statistical application of an MP1 matrix is next. This will be followed by the analogue formulation of the Moore-Penrose conditions for an arbitrary ring, R.

---

1   E. H. Moore, "On The Reciprocal Of The General Algebraic Matrix", <u>Bull. Am. Math. Soc.</u>, 26, 394–395.

2   R. Penrose, "A Generalized Inverse For Matrices", <u>Proc. Camb. Philos. Soc.</u>, 51, 406–413.

IB. REMARKS ON THE ALGEBRAIC TOPICS ORGANIZATION

For an arbitrary ring R the analogue of the Moore-Penrose conditions are:

$$axa = a \qquad (1)$$
$$xax = x \qquad (2)$$
$$(xa)^* = xa \qquad (3)$$
$$(ax)^* = ax \qquad (4)$$

where "*" denotes an involution on R. An involution on R is a mapping *:R $\rightarrow$ R which satisfies:

(i) $(a^*)^* = a$ for all a in R,
(ii) $(a + b)^* = a + b$ for all a,b in R
(iii) $(ab)^* = b^*a^*$ for all a,b in R.

Obviously, on a ring R, the involution mapping "*" mimics the conjugate transpose operator on a matrix ring $M_n(R)$. Clearly, the conjugate-transpose operator is itself a particular involution on matrix rings. To avoid ambiguity, an arbitrary involution on matrix rings will be clearly identified.

An arbitrary ring R which satisfies the first Moore-Penrose condition will be called MP1; if it satisfies the second Moore-Penrose condition then it will be called MP2, etc.

The objective of this algebra book is to describe commutative and noncommutative rings R which satisfy different Moore-Penrose conditions. *If a ring R is MP1, then it is also well-known as a Von Neumann regular ring.* Any MP1 ring is MP2, but it will be demonstrated that a MP2 ring is not necessarily MP1.
A most interesting aspect of MP2 rings (which produced many theoretical observations) was constructing an example of such a ring which is not MP1. In each case, if R is MPk (k = 1, 2, 3, 4), then it will be answered whether or not $M_n(R)$ is necessarily MPk. This entire process commences with the analyses of the MP1 and MP2 cases.

II. A CHARACTERIZATION OF MP1 (VON NEUMANN REGULAR) RINGS

IIA. PROPERTIES OF MP1 RINGS

In this section are listed some properties of MP1 rings. Many of these properties strongly characterize the structure of MP1 rings and illuminate their inner structures; moreover, some analogues or comparisons of MP1 properties to an MP2 ring will be made in the next chapter.

THEOREM II.1: In an arbitrary ring R, the following are true:
  (1)   R is MP1.
  (2)   Every principal right ideal is generated by an idempotent.
  (3)   Every finitely generated right ideal is a direct summand.

THEOREM II.2: In a commutative MP1 ring with unit every prime ideal is maximal.
PROOF: Let P be any prime ideal, with a $\notin$ P. Then ax - 1 $\in$ P since (ax-1)a $\in$ P; hence, 1 $\in$ P + aR. Thus P is maximal.

EXAMPLE II. 3: Consider F[[X]], the ring of formal power series containing elements of the form $a(x) = a_o + a_1x + a_2x^2 \ldots\ldots$ over the field F. It is well known that F[[X]] is local with unique maximal ideal M = XR.

According to K. R. Goodearl two necessary and sufficient conditions[3] for an associative ring with unit to be MP1 are:
  (1)   Every principal left ideal is generated by an idempotent.

---

3   K. R. Goodearl, Von Neumann Regular Rings (Pitmann Publishing Ltd.: London), 1973, 1.

(2) Every finitely generated left ideal is generated by an idempotent. However, S.S. Page in his paper "Regular Rings Are Very Regular"[4] points out that if "an idempotent" is replaced by "idempotents" in equation (2) above then the MP1 property can be asserted as well.

EXAMPLES OF MP1 RINGS:

EXAMPLE II. 4: An integral domain with unit 1 is MP1 provided that it is a division ring. Thus the rationals, Q, the reals, R, and the quarterions, H are MP1 as well as $Z/(p)$, where p is a prime integer.

EXAMPLE II. 5: $M_n(F)$, where F is field, is a noncommutative MP1 ring (n $\geq$ 2).

EXAMPLE II. 6: $Hom_D(V,V)$ is an MP1 ring where D is a division ring and V is a finite dimensional vector space over D (V is a D module).

EXAMPLE II. 7: If R is semisimple artinian then R is MP1.

EXAMPLE II. 8: A McCoy example[5] that is not a direct sum of fields or matrix rings over division rings:

Let S be the set of all subsets of a given nonempty set A, including the empty set $\varnothing$ and the entire set A. If a,b are in S, define ab to be a $\cap$ b, the intersection of a and b. Also, a + b is defined to be the symmetric difference (a - b) $\cup$ (b - a). Then S is a ring with unity. Obviously, S is MP1 since every element is idempotent. Also, nonzero idempotents are orthogonal if and only if they correspond to disjoint subsets of A. Note that S is not a direct sum of fields since different sets may have the same intersection with a given set thus violating uniqueness of inverses.

THEOREM II. 9: R is MP1 if and only if $M_n(R)$ is MP1.

A brief proof of this assertion can be found in Goodearl's cited work.

4   S. S. Page, "Regular Rings Are Very Regular", <u>Canad. Math. Bull.</u>, 25, 1982, 118.

5   Neal H. McCoy, <u>The Theory Of Rings</u> (The MacMillan Co.: New York), 7.

IIB. INTEGRAL EXTENSION OF AN MP1 RING

The question arises if an integral extension of an MP1 ring is MP1. The next section investigates under what conditions this proposition is true. It shall be shown that under certain basic restrictions on prime ideals of an MP1 ring this preserves the MP1 property in the integral extension. This shall be preceded by a brief review of the notion of integral dependence.

DEFINITION II. 10: Let R be a subring of S both sharing the same unity. Then S is integral over R if every element s of S satisfies an equation of the form:

$$x^n + a_{n-1}x^{n-1} + \ldots + a_1x + a_0 = 0. \ (a_i \in R), \text{ n a positive integer.}$$

The following theorems are taken from the classic paper on integral dependence by I. S. Cohen and A. Seidenberg.[6] It is assumed that R and S are commutative rings such that S contains, and has the same identity element as, R.

DEFINTION II. 11: If P is a prime ideal in R and Q is a prime ideal in S such that Q ∩ R = P, then Q lies over or contracts to P.

THEOREM II. 12: Let S be integral over R, and let the prime ideal Q in S lie over the prime ideal P in R, that is, Q ∩ R = P. Then P is maximal if and only if Q is maximal.

Next, the "lying-over" theorem:

THEOREM II. 13: Let S be integral over R. Then for every prime ideal P in R there exists a prime ideal Q in S lying over P.

This is succeeded by the "going-up" theorem:

THEOREM II. 14: Let S be integral over R. Let P be a prime ideal in R containing the ideal A. If U is an ideal in S such that U ∩ R = A, then there exists a prime ideal in S containing U and lying over P.

---

6   I. S. Cohen and A. Seidenberg, "Prime Ideals And Integral Dependence", <u>Bull. Amer. Math. Soc.</u>, 52, 1946, 252–261.

Unlike the "going-up" theorem, the "going-down" theorem requires assumptions on the zero-divisors of R and S.

THEOREM II. 15: Let R be an integral domain integrally closed in its quotient field (the only elements in the quotient field integral over R lie in R), S a ring integral over R, with none of its zero-divisors in R. Then if P and P′ are prime ideals in R with P′ $\subset$ P, then for every prime ideal Q in S lying over P there exists a prime ideal D, contained in S, and lying over P′.
At this interval, the main result draws near. But first,

THEOREM II. 16: Let S be an integral extension of R. If R is MP1, then any prime ideal Q in S is a maximal ideal in S.
PROOF: Let P be the contraction of the prime ideal of Q to R. Since R is MP1, R/P is a division ring (any prime MP1 ring is a division ring). S/Q is an integral domain and is an integral extension of R/P. By the "lying-over" theorem, S/Q must be a division ring and hence, Q is a maximal ideal in S.
Finally,

THEOREM II. 17: An integral extension of an MP1 ring with no nonzero nilpotent elements is MP1.
PROOF: Let S be the integral extension of the MP1 ring R. If Q is a prime ideal of S, then Q is maximal by the above theorem. Hence, since S has no nonzero nilpotent elements, S is MP1.[7]
Now suppose the ring R is contained in the ring S both with unity. Assume all elements in S which are integral over R is MP1. Does this imply that R is MP1? No. Consider the fact that the ring of integers Z is integrally closed in Q the ring of rationals. All elements of Q integral over Z lies in Z.

---

7   K. R. Goodearl, Von Neumann Regular Rings (Pitmann Publishing Ltd.: London), 1973, 13.

III. A CHARACTERIZATIONS OF MP2 RINGS

IIIA. PROPERTIES OF MP2 RINGS

After examining some properties which characterize MP1 rings, the analogue of these properties and others will be examined for MP2 rings at least to see which ones remain true and illuminate the structure of MP2 rings; consider the following definition:

DEFINITION: A ring R is MP2 if for each nonzero element a in R there exists a nonzero element x in R such that $xax = x$.

LEMMA: Let R be an associative ring with unity. If every nonzero principal left ideal of R contains a nonzero idempotent, then R is MP2.

PROOF: Let a be a nonzero element of R. Then Ra is the principal left ideal generated by a. By hypothesis, Ra contains a nonzero idempotent $e = e^2$. For some r in R, $e = ra$. Set $x = er$.

Note: If $er = 0$ then $0 = era = e^2 = e = 0$ which contradicts the fact that e is a nonzero idempotent. Then $(er)a(er) = e(ra)(er) = eeer = er$ since e is idempotent and $e = ra$. Hence, $x = er \neq 0$ and $xax = x$. Thus, R is MP2.

Volume II of An Introduction To Moore-Penrose rings shall contain more structural theorems that insure a given ring is MP2.

EXAMPLES OF MP2 RINGS:

EXAMPLE III. 1: If R is MP1, then R is MP2. PROOF: Let $a \in R$; since R is MP1 there exists an x in R such that $axa = a$. Set $y = xax$ then $yay = y$ and R is MP2.

EXAMPLE III. 2: If F is a field then $M_n(F)$ (n a positive integer) is MP2 since $M_n(F)$ is MP1.

EXAMPLE III. 3: Let $R = \oplus \sum F_i$, where F_i is a field, then R is MP2.

EXAMPLE III. 4: $Z/(n)$, where n is square-free, is MP2.

IIIB. MATRIX RING OVER AN MP2 RING

THEOREM III. 5: Let R be a ring with identity. Then $M_n(R)$ is MP2 provided that R is MP2.

THEOREM III. 6: Let R be an associative ring with identity. If $M_n(R)$ is MP2, then R is MP2.

Both these results will be proven in detail in Volume II.
COROLLARY III. 7: If R is a direct sum of fields, then R is MP2.

PROOF: If R is a direct sum of fields, then R is certainly MP1; hence, $M_n(R)$ is MP1 and thereby MP2. Hence, R is MP2.

If a ring R is commutative, Noetherian then the structure of R decomposes nicely into a direct sum of MP1 rings if it is MP2. This decomposition will be explored further in the next volume.

A great deal of tedious effort and rigorous research has gone into the task of finding an example of an MP2 ring that is not MP1. It is immediately obvious that if R is MP1 then it is certainly MP2 as demonstrated earlier. Hence all MP1 rings are contained in the larger class of MP2 rings. A seemingly innocuous question: "Is this containment of MP1 rings inside MP2 rings proper?" Though this question has been an intractable one to answer either way, it has led to many good theorems and observations about the characteristics of both MP2 and MP1 rings. Given below is an excursion through Robert F. Shanny's paper on "Regular Endomorphism Rings Of Free Modules"[8] which surprisingly provide some subtle insight on necessary traits of an MP2 matrix ring not to be MP1 if such a matrix ring exists. It is an established fact that if R is

---

8   Robert F. Shanny, "Regular Endomorphism Rings Of Free Modules", J. London Math. Society, 2, 1971, 353–354.

an MP1 ring and F is a free R-module with finite basis, then $\text{Hom}_R(F,F)$ is an MP1 ring. Shanny removes the restriction that the basis be finite but shows that $\text{Hom}_R(F,F)$ is MP1 if and only if R is semisimple Artinian.

Since an MP2 ring is semisimple, if there is an MP2 matrix ring which is not MP1, then surely it must not be Artinian for else it would underscore the sufficiency of Shanny's observation. Thus the class of infinte matrix rings come into an interesting focus as perhaps fertile grounds for discovering if there is an MP2 ring not MP1.

THEOREM III. 8: Assume R has a unit and no nonzero nilpotent elements. R is MP1 if and only if for all proper ideals I, R/I is MP2.

PROOF: Since R is MP1, any homomorphic image is MP1. Hence, for all proper ideals I, R/I is MP2. For any prime ideal P, R/P is MP2 also. Hence, R/P is an MP2 domain. Thus, R/P is a field and P is maximal. By Edward T. Wong's results, R is MP1.[9]

Indeed, it is curious to note that if there is an MP2 ring that is not MP1, then a candidate may be an MP2 ring with no nonzero nilpotent elements and a proper homomorphic image that is not MP2.

---

9   Edward T. Wong, "Regular Rings And Integral Extension Of A Regular Ring", AMS, Vol. 33, 2, 1972.

IV. A CHARACTERIZATIONS OF MP3 RINGS

IVA. RINGS WITH INVOLUTION AND AN MP3 RING DEFINED

DEFINITION IV. 1: A ring R is said to have involution "*", where *: R → R, if it satisfies:

   (1) $(a^*)^* = a$ for all a in R

   (2) $(a + b)^* = a^* + b^*$ for all a,b in R

   (3) $(ab)^* = b^*a^*$ for all a,b in R

There are many special sets in a ring R with involution; in particular, $\{x \in R \mid x^* = x\}$ shall be called the set of symmetric elements in R and denoted by the letter S.

DEFINITION IV. 2: A ring R with involution "*' is called MP3 if for every nonzero element a in R,

$$Ra \cap S \neq 0.$$

In other words, a ring R with involution is called MP3 if the left ideal generated by each of its nonzero elements contains a nonzero symmetric element. Now if the involution "*" is taken to be the opertion of matrix transpose, then this definition agrees with the matrix definition of the MP3 property.

Note: The properties of an MP3 ring with involution "*" seems to be pretty much bound to the character of its set of symmetric elements S. When algebraic conditions are imposed on S, the structure of R is largely determined. This perspective will be reinforced later.

PROPOSITION IV. 3: If R is a ring with identity 1 and involution *, then is the identity symmetric?

PROOF: Let r be an arbitrary nonzero element of R. Let 1 be the identity of R. Now it follows that $r^* = (r \cdot 1)^* = (1 \cdot r)^* = 1^* \cdot r^* = r^* \cdot 1^*$. Since 1 is the identity of R, it follows that $r^* = 1 \cdot r^* = r^* \cdot 1$. So $1^* \cdot r^* = 1 \cdot r^* = r^* \cdot 1$ implies $(1^* - 1)r^* = 0$ or $(1 - 1^*)r = 0$ for all r in R. Similarly, $r(1^* - 1) = 0$ for all r in R. In particular, both equations hold true for $r = 1$. Then $1 - 1^* = 0$ or $1 = 1^*$. Hence, 1 is symmetric.

PROPOSITION IV. 4: If R is a ring with involution * then $0^* = 0$.

PROOF: Let a be an arbitrary nonzero element of R. Then $0 = a^* - a^* = (a - a)^* = (0)^* = 0^*$.

IVB. EXAMPLES OF RINGS WITH INVOLUTION AND MP3 RINGS

DEFINITION IV. 5: A ring R with involution * and symmetric elements $S = \{a \in R | a^* = a\}$ is MP3 if $Ra \cap S \neq 0$ for all a in R.

EXAMPLE IV. 6: Let C = the field of complex numbers with involution "*" equal to conjugation represented by "–", i. e., if $z = a + bi$, then $z^* = \bar{z} = a - bi$. Note that complex conjugation is a field automorphism of order 2. In fact, any field with an automorphism of order 2 may be classified as a ring with involution with involution being the particular automorphism. Note that in the case of C, the set of symmetric elements $S = \mathcal{R}$ the set of real numbers. Now consider the Galois group $G(Q(\sqrt{2})/Q)$ which is isomorphic to $Z_2 = \{I_{Q(\sqrt{2})}, \varphi\}$. Set $* = \varphi$; then $S = Q$ (fixed field) under φ. Both of these examples are trivially MP3. The second example may be generalized by the following assertion:

THEOREM IV. 7: Let E be a normal extension of F and $\mathcal{G}(E/F)$ denote the Galois group of E over F. If $\mathcal{G}(E/F) = \{\sigma_0, \sigma_1\}$ where σ_0 is the identity automorphism on E, then E is an MP3 ring with involution σ_1. Further, F is the set of symmetric elements of E under σ_1.

PROOF: Since $\mathcal{G}(E/F)$ is a multiplicative group, obviously $\sigma_1^2 = \sigma_0$ the identity automorphism. Hence, if $\alpha \in E$, then $\sigma_1(\sigma_1(\alpha)) = \alpha$. Also, given $\alpha, \beta \in E$ then

$\sigma_1(\alpha+\beta) = \sigma_1(\alpha) + \sigma_1(\beta)$ since σ_1 is an automorphism of E and preserves its additive structure. Now $\sigma_1(\alpha - \beta) = \sigma_1(\alpha) - \sigma_1(\beta) = \sigma_1(\beta) - \sigma_1(\alpha)$. Hence, σ_1 is an involution. Naturally, $\sigma_1(\alpha) = \alpha$ if and only if $\alpha \in F$. Thus, $F = S$, the set of symmetric elements of E under σ_1. To show E is MP3, let $\alpha \in E$ and $\alpha \neq 0$. Set $\gamma = \sigma_1(\alpha) - \alpha$. Then $\gamma \neq 0$ since $\alpha \neq 0$. Accordingly, $\sigma_1(\sigma_1(\alpha) \cdot \alpha) = \alpha \cdot \sigma_1(\alpha) = \sigma_1(\alpha) \cdot \alpha$. Hence, E is MP3.

To strengthen this result, consider the following theorem by I.N. Herstein[10] which requires a definition:

DEFINITION IV. 8: (opposite ring ^{op}) If R is a ring, then its opposite ring, R^{op}, is the ring having the same additive group as R but with multiplication "$\circ$" defined by $a \circ b = ba$.
If R is commutative, then $R \cong R^{op}$.

THEOREM IV. 9: Let R be a semi-prime ring in which every symmetric element $a \neq 0$ is invertible in R. Then R is
1. a division ring, or
2. the direct sum of a division ring and its opposite, relative to the exchange involution $(x, y)^* = (y, x)$, or
3. the 2 x 2 matrices over a field, relative to the symplectic involution, namely $\begin{pmatrix} \alpha & \beta \\ \gamma & \delta \end{pmatrix}^* = \begin{pmatrix} \delta & -\beta \\ -\gamma & \alpha \end{pmatrix}$

Using Herstein's result, note that the previous two examples are semiprime rings in which every symmetric element is invertible. In both cases, R was a division ring.

EXAMPLE IV. 10: Take a direct sum of copies of $\mathcal{C}$, the field of complex numbers as shown below:

$$\oplus \sum_{i \in I} \mathcal{C}_i \text{ where } S = \oplus \sum_{i \in I} \mathcal{R}_i \text{, direct copies of the real numbers. The involu-}$$

tion is complex conjugation applied separately to each direct summand.

EXAMPLE IV. 11: Now consider $M_n(\mathcal{C})$, if $A = (a_{ij})$ is an element of $M_n(\mathcal{C})$ then set $A^* = (\overline{a}_{ji})$. Then * is an involution on $M_n(\mathcal{C})$ where "*" is complex-conjugate transpose, i.e., $A^* = \overline{A}^T$. Now "*" can also be defined as ordinary

10   I.N. Herstein, <u>Rings With Involution</u> (University of Chicago Press: Chicago), 1976, 62.

transpose. When "*" is defined as the complex-conjugate transpose on $M_n(\mathcal{C})$, then the set of symmetric elements S under this involution are the the set of Hermitian matrices. In both cases, $M_n(\mathcal{C})$ is MP3. To verify this proposition, let $A \in M_n(\mathcal{C})$ be arbitrary $(A \neq 0)$; then $A^=A = I_R$, the projection (which is symmetric and idempotent) onto the column space of A. $A^=$, the pseudoinverse of A, certainly exists over the complex field $\mathcal{C}$. Thus for any nonzero $A \in M_n(\mathcal{C})$, $\exists X \in M_n(\mathcal{C})$ $(X \neq 0)$ such that XA is symmetric.

DEFINITION IV. 12: A ring R with involution "*" is called *richly MP3* if $\forall a \in R$, $a^*a = 0 \Rightarrow a = 0$.

EXAMPLE IV. 13: Let V be the vector space of polynomials over $\mathcal{C}$ with inner product

$\prec, \succ : V \times V \Rightarrow \mathcal{C}$ given by $\prec f, g \succ = \int_0^1 f(t)\bar{g}(t)dt$ where the polynomial f has the

form $\sum a_k x^k$. Let $f^* = \sum \bar{a}_k x^k$ where $a_k \in \mathcal{C}$ and "$^-$" denotes complex con-

jugation (so $f^* = \bar{f}$). Now define M_f by $M_f(g) = fg$. Then $\prec M_f(g), h \succ = \prec fg, h \succ$

$= \int_0^1 fg\bar{h}dt = \int_0^1 g(\overline{\bar{f}h})dt = \prec g, f^*h \succ = \prec g, M_{f^*}(h) \succ$. Thus, $(M_f)^* = M\bar{f}$.

If L(V) is the algebra of linear operators on a vector space V, then $A(V) = \{M_f |$ $f \in \mathcal{C}[x]\}$ is a subring of L(V) with involution "*" being $(M_f)^* = M\bar{f}$. The set of symmetric elements, S, is the set of linear operators M_f, where f is a polynomial with real coefficients. Note A(V) is richly MP3 since it is an integral domain. Note also that the only invertible elements of S are nonzero constant polynomial operators.

EXAMPLE IV. 14: Take any finite dimensional inner product space over $\mathcal{R}$, the reals or $\mathcal{C}$, the complex number field. Then the ring of linear operators $\mathcal{A}$ on it with "*" involution defined as the adjoint operaton is well-defined, i.e.,

given T: $V \Rightarrow V$ then $\prec Ta, b \succ = \prec a, T^*b \succ$

For the reals numbers $\mathcal{R}$, and complex numbers $\mathcal{C}$, the symmetric elements are the symmetric and Hermitian operators respectively. The ring $\mathcal{A}$ is MP3 since any operator can be premultiplied by its pseudoinverse operator.

EXAMPLE IV. 15. Consider $\mathcal{C}[x]$ or $\mathcal{C}[[x]]$, the complex formal power series: let *:$f(x) \Rightarrow \bar{f}(x)$; then for $f \in \mathcal{C}[x]$, $f(x) = \sum_{k=0}^{n} a_k x^k$ and $f(x)^* = \sum_{k=0}^{n} \bar{a}_k x^k$ where $a_k \in \mathcal{C}$.
Also, for $g \in \mathcal{C}[[x]]$, if $g(x) = \sum_{0}^{\infty} b_n x^n$ then $g^*(x) = \sum_{0}^{\infty} \bar{b}_n x^n$ where $b_n \in \mathcal{C}$.

The set of symmetric elements for $\mathcal{C}[x$ is $S = R[x]$ and the set of symmetric elements for $\mathcal{C}[[x]]$ is $S = R[[x]]$ in the case of the formal power series. Since both rings are integral domains, they are richly MP3.

PROPOSITION IV. 16: If R is an integral domain with involution "*" then R is richly MP3.

PROOF: Let $a \in R$ ($a \neq 0$). Then $a^*a \in S$, the set of symmetric elements in R, since $(a^*a)^* = (a)^*(a^*)^* = a^*a$. Now $a^*a \neq 0$ since otherwise $a^*a = 0 \Rightarrow$ either $a = 0$ or $a^* = 0$. Either case contradicts the fact that $a \neq 0$. Hence, R is richly MP3.

IVC. MORE CHARACTERIZATIONS OF MP3 RINGS

Note: In the statements which follow, R is assumed associative with unit.
OBSERVATIONS:

PROPOSITION IV. 17: (a) If R is a commutative ring with involution "*" and S intersects all its proper ideals nontrivially, then R is MP3.
PROPOSITON IV. 18: (b) If R is a subdirectly irreducible ring with involution and S, the set of symmetric elements, then if S is an ideal, it follows that R is MP3.

DEFINITION IV. 19: If R is a ring with involution "*" and S is the set of symmetric elements then S is said to be *saturated with idempotents* if $\forall s \in S$, $s^2 = s$.

EXAMPLE IV. 20: Consider $M_2(D_2)$, the ring of 2 x 2 diagonal matrices over the field Z_2. Taking involution to be the matrix transpose, then some short computations show that $M_2(D_2)$ is saturated with idempotents. In general, $M_n(D_2)$, the ring of n x n diagonal matrices over the field Z_2 is saturated with idempotents.

The next result is from Charles Lanski's paper, *"Rings With Involution Which Are I-rings"*:

THEOREM IV. 21: Assume R is a ring with involution * which has no nilpotent elements and that for each s∈ S, the set of symmetric elements, there is a y ∈ R such that sys = s. Then R is MP1.

COROLLARIES:

COROLLARY IV. 22: If R is a simple ring with no nonzero nilpotent elements and involution "*" and S ⊆ Z(R) then R is MP1 and trivially MP3.

COROLLARY IV. 23: If R is a ring with involution "*" having no nonzero nilpotent elements and all elements of S are invertible, then R is MP1.

COROLLARY IV. 24: If R is a commutative, noetherian ring with involution "*", MP3, and S is saturated with idempotents, then R is MP1.

PROOF OF Corollary IV. 22 Since R is simple, Z(R) is a field and thus for each s in S, there exists an y such that sys − s. Hence, by Lanski's theorem, R is MP1. Also, R is trivially MP3 since for each nonzero a in R, a*a is non-zero.

PROOF OF Corollary IV. 23: Apparent.

PROOF OF Corollary IV. 24: R is an MP2 ring and since R is commutative, noetherian MP2, then R is a direct sum of fields. Hence, R is MP1.

PROPOSITION IV. 25: If R is a commutative ring with involution * then S is multiplicatively closed when all of its elements are invertible. If R is also MP3 then R has no proper left ideals. All elements of R are left invertible; hence R is a field.

PROPOSITION IV. 26: Let R be a ring with identity and involution *. If all the elements of s are invertible and all ideals I are such that $I^* = I$, then every homomorphic image of R is MP1.

PROOF: Straightforward. R/I is 2-torsion free and S has no nonzero nilpotent elements. By Lanski's result, R is MP1. Then with the given hypotheses, each homomorphic image is MP1 induced.

PROPOSITION IV. 27: Let R be a ring with involution *; if R is prime and has no nonzero nilpotent elements, then R is MP3.

PROOF: By the hypothesis, R has no nonzero divisors of zero. Hence, given a $\in$ R, a^*a is symmetric and R is MP3.

THEOREM IV. 28: A direct sum of MP3 rings is MP3.

PROOF: Let R_1, ..., R_n be MP3 rings with $*^i$ being the involution on R_i respectively. Let $R = \oplus \sum_{i=1}^{n} R_i$ and define the involution *: R $\to$ R in this manner: r $\in$ R has the representation $r = (r_1, ..., r_n)$ where $r_i \in R_i$ and $r^* = (r_1^{*1}, ..., r_n^{*n})$. Let S_i be the set of symmetric elements of R_i. The set S of symmetric elements of R is represented then as $(S_1, ..., S_n)$. Now to show R is MP3, let $a \in R$, $a \neq 0$ so $a = (a_1, ..., a_n)$ where $a_i \in R_i$ and not all a_i are zero. Since each R_i is MP3, $\exists r_i \in R_i$, $r_i \neq 0$ such that $r_i a_i \in S_i$, not all $r_i a_i$ equal zero, and $(r_i a_i)^{*i} = r_i a_i$. Now set $r = (r_1, ..., r_n)$. Then $ra = (r_1 a_1, ..., r_n a_n) = ((r_1 a_1)^{*1}, ..., (r_n a_n)^{*n}) = (ra)^* \in S$ and $ra \neq 0$. Hence, R is MP3.

THEOREM IV. 29: A direct summand of an MP3 ring is MP3 under the induced involution.

PROOF: Let R be an MP3 ring with involution * and T a direct summand of R. Then $R = T \oplus V$ where $T \cap V = \{0\}$. If S is the set of symmetric elements of R then $S \cap T$ is the set of symmetric elements of T and $S \cap V$ is the set of symmetric elements of V. Assume $t \in T$, $t \neq 0$. Thereby $t' = (t, 0) \in R$. Since R is MP3, $\exists r \in R$, $r \neq 0$ such that $rt' \in S$ and $rt' \neq 0$.
Using the representation $r = (\rho, v)$ then $\rho t \in S \cap T$, $\rho t \neq 0$ since otherwise $rt' = 0$. So $(\rho t)^{*T} = \rho t$ where $*^T$ is the involution on T induced by restricting the involution *: R $\to$ R to T. Hence, T is MP3.

IVD. LOCALIZATION OF AN MP3 RING

THEOREM IV. 30: Let R be a ring with identity and involution *. Further let S be the set of symmetric elements of R and set $S' = S - \{0\}$. Assume F is a saturated multiplicatively closed subset of R such that $S' \subseteq F$. If R is MP3, then R_F is MP3.

PROOF: Let a/b be an arbitrary nonzero element of R_F. Since $a,b \in R$ and R is MP3, there exists
$r, p \in R$ such that $ra \in S$, $ra \neq 0$ and $pb \in S$, $pb \neq 0$. Se t = r/p. Then t(a/b) = ra/pb = (ra)*/(pb)* lies in S_F, the set of symmetric elements of R_F. Hence, R_F is MP3.

THEOREM IV. 31: Let R be a ring with involution "*", S its set of symmetric elements and T a ring with involution "⁺". Let $\varphi : R \to T$ be a ring homomorphism such that for each r in R,
$\varphi(r^*) = \varphi(r)^{+}$ and $S \cap \ker\varphi = \{0\}$. If R is MP3, then the image of R in T is MP3.

PROOF: Let a' be an element of $\varphi(R) \leq T$, $a' \neq 0$. Then there exists an a in R, $a \neq 0$, such that $\varphi(a) = a'$. But R MP3 implies there exists an r in R such that ra lies in S, the symmetric elements of R and $ra \neq 0$. Then $\varphi(ra) = \varphi(r) \varphi(a)$ $= \varphi((ra)^*) = \varphi(a^*r^*) = \varphi(a^*)\varphi(r^*) = \varphi(a)^{+} \varphi(r)^{+} = (\varphi(r) \varphi(a))^{+}$. Hence, $\varphi(r)$ $\varphi(a) = \varphi(r)a' = (\varphi(r) a')^{+}$. Now $\varphi(r)a' \neq 0$ since otherwise $\varphi(r)a' = 0$ implies $\varphi(a^*r^*) = \varphi(ra) = 0$ which implies ra lies in the $\ker\varphi$, and this implies that ra = 0. Contradiction!! Therefore, $\varphi(r)a' \neq 0$ and $\varphi(R)$ is MP3 under the involution "⁺".

COROLLARY 1V. 32: If R is richly MP3 then any homomorphic image of R is MP3 with the induced involution.

EXAMPLE IV. 33: Let R be the ring $\left\{ \begin{bmatrix} x & 0 & 0 \\ 0 & y & 0 \\ 0 & 0 & z \end{bmatrix} \mid x,y,z \in R \right.$, the field of real numbers $\left. \right\}$. Now set $A = \begin{bmatrix} a & 0 & 0 \\ 0 & b & 0 \\ 0 & 0 & c \end{bmatrix}$ and let the homomorphism φ be left multiplication by $\begin{bmatrix} 1 & 0 & 0 \\ 0 & 1 & 0 \\ 0 & 0 & 0 \end{bmatrix}$. If $A \neq 0$ then $\varphi(A) = \begin{bmatrix} a & 0 & 0 \\ 0 & b & 0 \\ 0 & 0 & 0 \end{bmatrix} \neq 0$. Let the involution * on R be matrix transpose. Then A* = A so that all elements of R are

symmetric. Hence, R is MP3. In fact, R is richly MP3. To see this, let B =

$\begin{bmatrix} b_1 & 0 & 0 \\ 0 & b_2 & 0 \\ 0 & 0 & b_3 \end{bmatrix}$; then B*B = $\begin{bmatrix} b_1^2 & 0 & 0 \\ 0 & b_2^2 & 0 \\ 0 & 0 & b_3^2 \end{bmatrix}$ = 0 if and only if $b_1 = b_2 = b_3 = 0$ if and

only if B = 0. Now ker$\varphi \supseteq \{ \begin{bmatrix} 0 & 0 & 0 \\ 0 & 0 & 0 \\ 0 & 0 & g \end{bmatrix} \mid g \in R$, the field of real numbers$\}$

and ker $\varphi \cap R \neq 0$. In addition, $\varphi(R) = \{ \begin{bmatrix} u & 0 & 0 \\ 0 & v & 0 \\ 0 & 0 & 0 \end{bmatrix} \mid u,v \in R$, the field of real

numbers$\}$. Suppose the involution $\star$ on $\varphi(R)$ is matrix transpose also. If $C \in$

R, $C \neq 0$, then $\varphi(C^*) = \varphi(C^T) = \varphi(C) \star = \varphi(C)^T$. (*Note: Superscript "T" is the*

standard notation for matrix transpose). Hence, under matrix tranpose, $\varphi(R)$

is MP3. However, let $\star$, the involution on $\varphi(R)$ be interchanging the (1, 1) and

(2, 2) entries of an element of $\varphi(R)$. Now if D = $\begin{bmatrix} 0 & 0 & 0 \\ 0 & d & 0 \\ 0 & 0 & 0 \end{bmatrix} \neq 0$ then D$\star$ = D

if and only if d = 0 if and only if D = 0. Hence, $\varphi(R)$ under the diagonal-entry

exchange involution $\star$ is not MP3.

THEOREM IV. 34: Let R be an MP3 ring with identity and involution "*".

Then R[x] is MP3 with involution defined by: given f $\in$ R[x] where f =

$\sum_{k=0}^{n} a_k x^k$ then f* = $\sum_{k=0}^{n} (a_k) * x^k$.

PROOF: Let $g \in R[x]$, $g \neq 0$ where without loss of generality, g has a nonzero constant term. Too see this, suppose $h \in R[x]$ has the form $a_m x^m + a_{m+1} x^{m+1} + \ldots + a_n x^n$. Then $h = x^m (a_m + a_{m+1} x + \ldots + a_n x^{n-m}) = (a_m + a_{m+1} x + \ldots + a_n x^{n-m}) x^m$. If $R[x]$ is MP3 then there exist an p in $R[x]$ ($p \neq 0$) such that ph is nonzero symmetric. Then $(ph)^* = h^* p^* = (a_m + a_{m+1} x + \ldots + a_n x^{n-m})^* (x^m)^* p^* = (x^m)^* (a_m + a_{m+1} x + \ldots + a_n x^{n-m})^* p^* = x^m (a_m + a_{m+1} x + \ldots + a_n x^{n-m})^* p^* = x^m p (a_m + a_{m+1} x + \ldots + a_n x^{n-m})$ since $(x^m)^* = 1^* x^m = x^m$ because $1 = (1^*)^* = (1 \cdot 1)^* = (1^*)^* \cdot 1^* = 1 \cdot 1^* = 1^*$. Hence, g may be assumed to have a nonzero constant term. So let $g = a_0 + a_1 x^1 + \ldots + a_n x^n$ where a_i lies in R for all i. Then $g^* = (a_0)^* + (a_1)^* x^1 + \ldots + (a_n)^* x^n$. Now $g^* g = (a_0)^* a_0 + [(a_1)^* a_0 + (a_0)^* a_1] x^1 + \ldots +$

$$[\, a_n^* a_0 + a_{n-1}^* a_1 + \ldots + a_1^* a_{n-1} + a_0^* a_n \,] x^n.$$

Since at least $(a_0)^* a_0 \neq 0$ and all the coefficient expressions of the positive powers of x are symmetric expressions, then $g^* g$ is symmetric and nonzero. Consequently, $R[x]$ is MP3.

COROLLARY IV. 35: If R is an integral domain with involution "*", then $R[x]$ is MP3.

PROOF: Since R is an integral domain, $a^* a = 0$ if and only if $a = 0$ which means that R is richly MP3. Now using the induced polynomial involution on $R[x]$, then by the theorem above, $R[x]$ is MP3.

IVE. MATRIX RING OVER AN MP3 RING

THEOREM IV. 36: If $M_n(R)$ is MP3 then R is MP3.

PROOF: Let $a \in R$, $a \neq 0$. Put $A = \begin{bmatrix} a & 0 & \ldots & 0 \\ 0 & 0 & \ldots & 0 \\ \cdot & & & \\ \cdot & & & \\ 0 & 0 & \ldots & 0 \end{bmatrix} \neq 0$ in $M_n(R)$. Then there

exists an X in $M_n(R)$, $X \neq 0$ such that $(XA)^* = XA \neq 0$. If $X = (x_{ij})$,

then $XA =$
$$\begin{bmatrix} x_1 & x_2 & \cdots & x_n \\ x_2 & x_2 & \cdots & x_n \\ \cdot & & & \\ \cdot & & & \\ x_n & x_n & \cdots & x_n \end{bmatrix} \begin{bmatrix} a & 0 & \cdots & 0 \\ 0 & 0 & \cdots & 0 \\ \cdot & & & \\ \cdot & & & \\ 0 & 0 & \cdots & 0 \end{bmatrix} =$$

$$\begin{bmatrix} x_{11}a & 0 & \cdots & 0 \\ x_{21}a & 0 & \cdots & 0 \\ \cdot & & & \\ \cdot & & & \\ x_{n1}a & 0 & \cdots & 0 \end{bmatrix} . \text{ Then } \begin{bmatrix} x_{11}a & 0 & \cdots & 0 \\ x_{21}a & 0 & \cdots & 0 \\ \cdot & & & \\ \cdot & & & \\ x_{n1}a & 0 & \cdots & 0 \end{bmatrix} = XA = (XA)^* =$$

$$\begin{bmatrix} (x_{11}a)^* & (x_{21}a)^* & \cdots & (x_{n1}a)^* \\ 0 & \cdots & \cdots & 0 \\ \cdot & & & \\ \cdot & & & \\ 0 & \cdots & \cdots & 0 \end{bmatrix} . \text{ Hence, } x_{i1}a = 0, \text{ for all } i > 1. \text{ Since } XA \neq 0,$$

must have

$x_{11}a \neq 0$. Thus, $x_{11}a = (x_{11}a)^*$ and thereby R is MP3.

THEOREM IV. 37: Assume R is a ring with identity and involution * and suppose R contains no nonzero nilpotent elements. Then R is MP3 if and only if $M_n(R)$ is MP3.

PROOF: Suppose R is MP3 and let $A \in M_n(R)$, $A \neq 0$. Let a_{ij} be the first nonzero entry of A. Without loss of generality, assume $i = 1$ since A is row equivalent to the matrix resulting from switching the ith and 1st rows of A.

Hence, A =
$$\begin{bmatrix} 0 \ldots 0 \ a_{1j} \ a_{1,j+1} \cdots a_{1n} \\ \\ \\ \\ \\ \end{bmatrix}$$
and set $E_1 =$
$$\begin{bmatrix} 1\ 0 \ldots \ldots 0 \\ 0 \ldots \ldots \ldots 0 \\ \cdot \\ \cdot \\ 0 \ldots \ldots 0 \end{bmatrix}.$$

Then $E_1 A =$
$$\begin{bmatrix} 0 \ldots 0 a_{1j} \ a_{1,j+1} \cdots a_{1n} \\ 0 \ldots \ldots \ldots \ldots \ldots 0 \\ \cdot \\ \cdot \\ 0 \ldots \ldots \ldots \ldots 0 \end{bmatrix}.$$
Since R is MP3, $\exists r_j, \ldots, r_n$ in R

such that $r_j a_{1j} = s_j, \ldots, r_n a_{1n} = s_n$ where $s_j, s_{j+1}, \ldots, s_n$ are all nonzero symmetric elements in S. Now

set $P =$
$$\begin{bmatrix} r_j\ 0 \ldots \ldots 0 \\ 0 \ldots \ldots 0 \\ \cdot \\ \cdot \\ 0 \ldots \ldots 0 \end{bmatrix}.$$
Then $PE_1 A =$
$$\begin{bmatrix} 0 \ldots 0\ s_j \ r_j a_{1,j+1} \cdots r_j a_{1n} \\ 0 \ldots \ldots \ldots \ldots \ldots 0 \\ \cdot \\ \cdot \\ 0 \ldots \ldots \ldots \ldots \ldots 0 \end{bmatrix}.$$

Now set $C = PE_1 A$;

then C^*C has $s_j^2 \neq 0$ in the (j,j) position since R has no nonzero nilpotent elements and C^* is the involution transpose of C. Hence, set $X = C^*PE_1$. Now X $\neq 0$ since otherwise $XA = C^*C = 0 \Rightarrow s_j^2 = 0$
which is a contradiction. Therefore, $M_n(R)$ is MP3.

Now assume $M_n(R)$ is MP3. Then by the previous theorem, R is MP3.

COROLLARY IV. 38: Assume R is a ring with identity and involution * and suppose that each element of S, the set of symmetric elements, is idempotent. Then R is MP3 if and only if $M_n(R)$ is MP3.

PROOF: A nonzero idempotent element cannot be nilpotent.

DEFINITION IV. 39: An associative ring R is called *formally real* if for any finite set of elements $\{ a_i \}$ $(1 \leq i \leq n < \infty)$ in R such that $\mathbf{a_1^2 + a_2^2 + \ldots + a_n^2}$ $= 0$ then $a_1 = a_2 = \ldots = a_n = 0$.

THEOREM IV. 40: Let R be an MP3 ring with involution "*"; also, let R be formally real and S saturated with idempotents. Then $M_n(R)$ is MP3 if for all a in R, $a*a = 0$ implies $a = 0$.

PROOF: Suppose for a nonzero $A \in M_n(R)$, $A*A = 0$. Then in particular, for $A = (a_{ij})$:

$$a_{11}*a_{11} + a_{21}*a_{21} + \ldots + a_{n1}*a_{n1} = 0$$

.

.

.

$$a_{1n}*a_{1n} + a_{2n}*a_{2n} + \ldots + a_{nn}*a_{nn} = 0$$

But $a_{ij}*a_{ij}$ is symmetric; hence, $(a_{ij}*a_{ij})^2 = a_{ij}*a_{ij}$. Then,

$$(a_{11}*a_{11})^2 + \ldots + (a_{n1}*a_{n1})^2 = 0$$

.

.

.

$$(a_{1n}*a_{1n})^2 + \ldots + (a_{nn}*a_{nn})^2 = 0$$

which implies $a_{11}*a_{11} = \ldots = a_{nn}*a_{nn} = 0$ since R is formally real. Thus, $\forall i,j$ $a_{ij} = 0$ by hypothesis and $A = 0$. Contradiction! $A \neq 0$. Hence, $M_n(R)$ is MP3.

COROLLARY IV. 40: If R is richly MP3 and formally real then $M_n(R)$ is richly MP3.

IVF. MP3, MP2 AND MP1 EQUIVALENCE

THEOREM IV. 41: Let R be a nontrivial ring with involution "*". Assume S, the set of symmetric elements, is nonzero and that each nonzero element of S is invertible. Further assume for any nonzero idempotent f there exists an element p in R such that pf is a nonzero symmetric element. Then the following statements are equivalent:

        (1) R is MP3.
        (2) R is MP2
        (3) R is MP1
        (4) R is a division ring.

PROOF: (1) $\Rightarrow$ (4) Assume R is MP3 and a lies in R (a $\neq$ 0). Since R is MP3, there exists an r in R such that ra lies S and ra $\neq$ 0. But ra symmetric implies that ra is invertible. Hence, there exists an y in R such that yra = 1. Set x = yr. Then x $\neq$ 0 since otherwise 0 = xa = 1. Contradiction! Hence, xa = 1 and R is a division ring Now, (4) $\Rightarrow$ (3) Since R is a division ring, R is MP1. (3) $\Rightarrow$ (2) R is MP1 and hence it is well-known to be MP2. (2) $\Rightarrow$ (1) Let a be an arbitrary element of R, a $\neq$ 0. Since R is MP2 there exists an r in R such that ra = e which is a nonzero idempotent element. By hypothesis there exists an p in R such that pe is nonzero symmetric. Set x = pr. Now
x $\neq$ 0 since otherwise xa = pe = 0. Contradiction! Hence, xa is nonzero symmetric and R is MP3.

IVG. INTEGRAL EXTENSION OF AN MP3 RING

THEOREM IV. 42: Let R be an associative MP3 ring with unity and involution "*". Let Q be an integral extension of R. Then Q is MP3 under the involution "*".

PROOF: Since Q is an integral extension of R, for arbitrary q $\in$ Q, there exists a positive integer n and elements a_{n-1}, a_{n-2}, ..., a_0 in R such that $q^n + a_{n-1}q^{n-1} + \ldots + a_1q + a_0 = 0$. Then adding $-a_0$ to both sides of this equation yields the following:

$$-a_0 = q^n + a_{n-1}q^{n-1} + \ldots + a_1q$$

Since R is MP3, there exists an r_0 in R ($r_0 \neq 0$) such that $r_0(-a_0) \neq 0$ and $r_0(-a_0)$ lies in S, the set of symmetric elements of R. Consequently,

$$r_0(q^n + a_{n-1}q^{n-1} + \ldots + a_1q) \in S$$

Since q is common to each term of the expression in the second factor of this equation, by the distributive law

$$r_0(q^{n-1} + a_{n-1}q^{n-2} + \ldots + a_1) q \text{ lies in S}$$

Now set $q' = r_0(q^{n-1} + a_{n-1}q^{n-2} + \ldots + a_1)$. Then $q' \neq 0$ (else $r_0(-a_0) = 0$) and $q'q$ lies in S. Hence, for arbitrary q $\in$ Q $\exists$ q' $\in$ Q such that $q'q$ is symmetric. Then Q is MP3 under the involution "*" of R.

IVH. MP3 MISCELLANEOUS RESULTS

DEFINITION IV. 43: Call a nonzero ideal J MP3-induced if J is MP3 in the sense that given a $\in$ J, a $\neq$ 0, then Ja $\cap$ S $\neq$ 0 where S is the set of symmetric elements in R.

THEOREM IV. 44: Assume R is a ring with involution "*" and identity. Suppose that for each nonzero ideal A, A* is an MP3-induced commutative subring of R. Then R is MP3.

PROOF: Let a be an arbitrary nonzero element of R and <a> denote the ideal generated by a. The ideal <a>* is a commutative MP3-induced subring of R. An arbitrary element of <a>* has the form

$\sum_{i=1}^{n} r_i a * s_i$. By hypothesis, since <a>* is MP3-induced, $\exists$ t $\in$ <a>* such that

$t(\sum_{i=1}^{n} r_i a * s_i) \in$ S and $t(\sum_{i=1}^{n} r_i a * s_i) \neq 0$. Since R has an identity and <a>* is

commutative, $t(\sum_{i=1}^{n} r_i a * s_i) = (\sum_{i=1}^{n} r_i a * s_i)t$.

But $s_i t \in$ <a>* and hence commutes with a* so that $(\sum_{i=1}^{n} r_i a * s_i)t = (\sum_{i=1}^{n} r_i s_i t)a*$

$= a*(\sum_{i=1}^{n} r_i s_i)t$

since $\sum_{i=1}^{n} r_i s_i t \in$ <a>* and commutes with a*. Hence, $t(\sum_{i=1}^{n} r_i a * s_i) = a*(\sum_{i=1}^{n} r_i s_i)t$.

But, $t(\sum_{i=1}^{n} r_i a * s_i) \in$ S $\Rightarrow a*(\sum_{i=1}^{n} r_i s_i)t \in$ S. Then $(a*(\sum_{i=1}^{n} r_i s_i)t)* = t*(\sum_{i=1}^{n} r_i s_i)*a \in$ S.

Now set $x = t* (\sum_{i=1}^{n} r_i s_i)*$. Hence, $x \neq 0$ since otherwise $a*(\sum_{i=1}^{n} r_i s_i)t = 0$ con-

trary to derivation.

Hence xa $\neq$ 0 and xa $\in$ S. Thus, R is MP3.

DEFINITION IV. 45: Now let R be a ring with involution "*". Let G be a mul-

tiplicative group. If x $\in \Gamma_R(G)$, (the group ring of R over G), then define x by x

$= \sum_{g} X_g g$ where $X_g \in$ R and where the involution

$*: \Gamma_R(G) \to \Gamma_R(G)$ is defined by $x* = \sum_{g} (X_g)* g^{-1}$ where g $\in$ G.

PROPOSITION IV. 46: If R is a MP3 domain and G is cyclic-finite, then $\Gamma_R(G)$ is MP3.

PROOF: Let $x = \sum_{i=0}^{n-1} X_i g^i$ where the order of G, $o(G) = n$. Now set $y = \sum_{i=0}^{n-1} (X_i)^* g^{n-i}$. Then it is readily computed that yx is nonzero symmetric. Hence, $\Gamma_R(G)$ is MP3.

PROPOSITION IV. 47: Let R be a ring with involution *. If for each a in R, $rad(Ra) \cap S \neq 0$, then R is MP3.

PROOF: Let a be an arbitrary nonzero element of R. If x lies in $Rad(Ra) \cap S$, then $x^* = x$ and $x^n = y \in Ra$. Say $x^n = r'a$. Then $(x^n)^* = (r'a)^* = (x^*)^n = x^n = r'a$.

V. CHARACTERIZATIONS
OF MP4 RINGS

The algebraic definition of an MP4 ring is a lettering transposition of the defining formula for an MP3 ring. Hence, most of the results for MP4 rings are exactly the same as in the MP3 case. They are enumerated specifically if the MP4 ring is being studied as a separate body of concepts independent of the MP3 case. Those sections of the MP4 literature analogous to the MP3 ring may be perused briefly, however. The mathematical results bridging the MP3 and MP4 rings are contained in Section VI, Chapter VII, "MP3 And MP4 Implications".

VA. MP4 MATRIX RING OVER A COMPLETE EQUICHARACTERISTIC LOCAL RING

DEFINITION V. 1: Given a ring R, its matrix ring $M_n(R)$ satisfies the MP4 property if for a given A in $M_n(R)$ there exists a nonzero X such that AX is a nonzero symmetric matrix.

In particular, it is noted that $M_2(R)$ is MP4 when R is a field F, or nontrivial local rings like Z_4 and $Z_2[[t]]/(t^2)$. One interesting feature which $Z_2[[t]]/(t^2)$ possesses, yet the former ring Z_4 does not, is the containment of a coefficient field. It shall be the primary objective of this section to determine which rings are MP4 and what the difference is for those rings which are not.

VB. MP4 RING DEFINED WITH INVOLUTION

Up to this point, a general matrix ring $M_n(R)$ has been defined as MP4 if it satisfies the fourth Moore-Penrose condition, i.e., $(AX)^* = AX, \forall A \in M_n(R)$; whereas the latter theorem provided an avenue to explore some interesting rings under this strictly formal definition, it is not enough to produce a wide variety of rings to study outside the structure of a matrix ring as in the MP1 and MP2 cases. What is desired is a condition on the elements of the ground ring R which will mimic the MP4 property of matrix ring elements. It turns out that rings with involution "*" are the ideal candidates to proceed with.

DEFINITION V. 2: A ring R is said to have involution "*", where *: $R \Rightarrow R$, if it satisfies:

$$(1)\ (a^*)^* = a \qquad \forall a \in R$$
$$(2)\ (a + b)^* = a^* + b^* \qquad \forall a,b \in R$$
$$(3)\ (ab)^* = b^*a^* \qquad \forall\ a,b \in R$$

There are many special sets in a ring R with involution; in particular, $\{x \in R \mid x^* = x\}$ shall be called the set of symmetric elements in R and denoted by the letter S.

DEFINITION V. 3: A ring R with involution "*' is called MP4 if for every nonzero element a in R, $aR \cap S \neq 0$.
In other words, a ring R with involution is called MP4 if the left ideal generated by each of its nonzero elements contains a nonzero symmetric element. Now if the involution "*" is taken to be the operation of matrix transpose, then this definition agrees with the matrix definition of the MP4 property.
Note: The properties of an MP4 ring with involution "" seems to be pretty much bound to the character of its set of symmetric elements S. When algebraic conditions are imposed on S, the structure of R is largely determined. This perspective will be reinforced later.*

VC. EXAMPLES OF RINGS WITH INVOLUTION AND MP4 RINGS

DEFINITION V. 4: A ring R with involution * and symmetric elements S = {a $\in$ R|a* = a} is MP4 if aR $\cap$ S≠ 0 for all elements a in R.

EXAMPLE V. 5: Let C = the field of complex numbers with involution "*" equal to conjugation represented by "−", i. e., if z = a + bi, then $z^* = \overline{z} = a - bi$. Note that complex conjugation is a field automorphism of order 2. In fact, any field with an automorphism of order 2 may be classified as a ring with involution with involution being the particular automorphism. Note that in the case of C, the set of symmetric elements S = $\mathcal{R}$ the set of real numbers.

Now consider the Galois group $G(Q(\sqrt{2})/Q) \approx Z_2 = \{i_{Q(\sqrt{2})}, \varphi\}$. Set $* = \varphi$; then $S = Q$ (fixed field) under φ. Both of these examples are trivially MP4. The second example may be generalized by the following assertion:

THEOREM V. 6: Let E be a normal extension of F and $\mathcal{G}(E/F)$ denote the Galois group of E over F. If $\mathcal{G}(E/F) = \{\sigma_0, \sigma_1\}$ where σ_0 is the identity automorphism on E, then E is an MP4 ring with involution σ_1. Further, F is the set of symmetric elements of E under σ_1.

PROOF: Since $\mathcal{G}(E/F)$ is a multiplicative group, obviously $\sigma_1^2 = \sigma_0$ the identity automorphism. Hence, if $\alpha \in E$, then $\sigma_1(\sigma_1(\alpha)) = \alpha$. Also, given $\alpha, \beta \in E$ then $\sigma_1(\alpha+\beta) = \sigma_1(\alpha) + \sigma_1(\beta)$ since σ_1 is an automorphism of E and preserves its additive structure. Now $\sigma_1(\alpha \cdot \beta) = \sigma_1(\alpha) \cdot \sigma_1(\beta) = \sigma_1(\beta) \cdot \sigma_1(\alpha)$. Hence, σ_1 is an involution.
Naturally, $\sigma_1(\alpha) = \alpha$ if and only if $\alpha \in F$. Thus, F = S, the set of symmetric elements of E under σ_1. To show E is MP4, let $\alpha \in E$ and $\alpha \neq 0$. Set $\gamma = \alpha \cdot \sigma_1(\alpha)$. Then $\gamma \neq 0$ since $\alpha \neq 0$. Accordingly, $\sigma_1(\alpha \cdot \sigma_1(\alpha)) = \sigma_1(\alpha) \cdot \alpha = \alpha \cdot \sigma_1(\alpha)$. Hence, E is MP4.
To strengthen this result, consider the following theorem by I.N. Herstein[11] which requires a definition:

DEFINITION V. 7: (opposite ring [op]) If R is a ring, then its opposite ring, R^{op}, is the ring having the same additive group as R but with multiplication "$\circ$" defined by a $\circ$ b = ba.
If R is commutative, then $R \cong R^{op}$.

THEOREM V. 8: Let R be a semi-prime ring in which every symmetric element a $\neq$ 0 is invertible in R. Then R is
  1. a division ring, or
  2. the direct sum of a division ring and its opposite, relative to the exchange involution $(x, y)^* = (y, x)$, or
  3. the 2 x 2 matrices over a field, relative to the symplectic involution, namely $\begin{pmatrix} \alpha & \beta \\ \gamma & \delta \end{pmatrix}^* = \begin{pmatrix} \delta & -\beta \\ -\gamma & \alpha \end{pmatrix}$

---

11   I.N. Herstein, <u>Rings With Involution</u> (University of Chicago Press: Chicago), 1976, 62.

Using Herstein's result, note that the previous two examples are semiprime rings in which every symmetric element is invertible. In both cases, R was a division ring.

EXAMPLE V. 9: Take a direct sum of copies of $\mathscr{C}$, the field of complex numbers as shown below: $\oplus \sum_{i\in I} \mathscr{C}_i$ where $S = \oplus \sum_{i\in I} \mathscr{R}_i$, direct copies of the real numbers. The involution is complex conjugation applied separately to each direct summand.

EXAMPLE V. 10: Now consider $M_n(\mathscr{C})$, if $A = (a_{ij})$ is an element of $M_n(\mathscr{C})$ then set $A^* = (\overline{a}_{ji})$. Then * is an involution on $M_n(\mathscr{C})$ where "*" is complex-conjugate transpose, i.e., $A^* = \overline{A}^T$. Now "*" can also be defined as ordinary transpose. When "*" is defined as the complex-conjugate transpose on $M_n(\mathscr{C})$, then the set of symmetric elements S under this involution are the the set of Hermitian matrices. In both cases, $M_n(\mathscr{C})$ is MP4. To verify this proposition, let $A \in M_n(\mathscr{C})$ be arbitrary $(A \neq 0)$; then $AA^{\mathcal{E}} = I_L$, the projection (which is symmetric and idempotent) onto the row space of A. $A^{\mathcal{E}}$, the pseudoinverse of A, certainly exists over the complex field $\mathscr{C}$. Thus for any nonzero $A \in M_n(\mathscr{C})$, $\exists X \in M_n(\mathscr{C})$ $(X \neq 0)$ such that AX is symmetric.

DEFINITION V. 11: A ring R with involution "*" is called *richly MP4* if $\forall a \in R$, $aa^* = 0 \Rightarrow a = 0$.

EXAMPLE VI. 14: Let V be the vector space of polynomials over $\mathscr{C}$ with inner product

$\prec , \succ : V \times V \Rightarrow \mathscr{C}$ given by $\prec f,g \succ = \int_0^1 f(t)\overline{g}(t)dt$ where the polynomial f has the form $\sum a_k x^k$. Let $f^* = \sum \overline{a}_k x^k$ where $a_k \in \mathscr{C}$ and "–" denotes complex conjugation (so $f^* = \overline{f}$). Now define M_f by $M_f(g) = fg$. Then $\prec M_f(g), h \succ =$

$\prec fg, h \succ = \int_0^1 fg\overline{h}dt = \int_0^1 g(\overline{fh})dt = \prec g, f^*h \succ = \prec g, M_{f^*}(h) \succ$. Thus, $(M_f)^* = M\overline{f}$.

If L(V) is the algebra of linear operators on a vector space V, then $A(V) = \{M_f |$ $f \in \mathscr{C}[x]\}$ is a subring of L(V) with involution "*" being $(M_f)^* = M\overline{f}$. The set of symmetric elements, S, is the set of linear operators M_f, where f is a polynomial with real coefficients. Note A(V) is richly MP4 since it is an integral domain. Note also that the only invertible elements of S are nonzero constant polynomial operators.

EXAMPLE V. 13: Take any finite dimensional inner product space over $\mathcal{R}$, the reals or $\mathcal{C}$, the complex number field. Then the ring of linear operators $\mathcal{A}$ on it with "*" involution defined as the adjoint operaton is well-defined, i.e.,

$$\text{given T: V} \Rightarrow \text{V} \qquad \text{then} \qquad \prec \text{Ta, b} \succ = \prec \text{a, T*b} \succ$$

For the reals numbers $\mathcal{R}$, and complex numbers $\mathcal{C}$, the symmetric elements are the symmetric and Hermitian operators respectively. The ring $\mathcal{A}$ is MP4 since any operator can be postmultiplied by its pseudoinverse operator.

EXAMPLE V. 14: Consider $\mathcal{C}[x]$ or $\mathcal{C}[[x]]$, the complex formal power series:

$*:f(x) \Rightarrow \bar{f}(x)$; then for $f \in \mathcal{C}[x]$, $f(x) = \sum_{k=0}^{n} a_k x^k$ and $f(x)^* = \sum_{k=0}^{n} \bar{a}_k x^k$ where

$a_k \in \mathcal{C}$. Also, for $g \in \mathcal{C}[[x]]$, if $g(x) = \sum_{0}^{\infty} b_n x^n$ then $g^*(x) = \sum_{0}^{\infty} \bar{b}_n x^n$ where

$b_n \in \mathcal{C}$. The set of symmetric elements for $\mathcal{C}[x$ is $S = R[x]$ and the set of sym-

metric elements for $\mathcal{C}[[x]]$ is $S = R[[x]]$ in the case of the formal power series.

Since both rings are integral domains, they are richly MP4.

PROPOSITION V. 15: If R is an integral domain with involution "*" then R is richly MP4.

PROOF: Let $a \in R$ ($a \neq 0$). Then aa* $\in$ S, the set of symmetric elements in R, since (aa*)* = (a*)*a* = aa*. Now aa* $\neq$ 0 since otherwise aa* = 0 $\Rightarrow$ either a = 0 or a* = 0. Either case contradicts the fact that a $\neq$ 0. Hence, R is richly MP4.

VD. MORE CHARACTERIZATIONS Of MP4 RINGS

Note: In the statements which follow, R is assumed associative with unit.

OBSERVATIONS:

PROPOSITION V. 16: If R is a commutative ring with involution "*" and S intersects all its proper ideals nontrivially, then R is MP4.

PROPOSITION VI. 17: If R is a subdirectly irreducible ring with involution and S, the set of symmetric elements, is an ideal, then R is MP4.

DEFINITION V. 18: If R is a ring with involution "*" and S is the set of symmetric elements then S is said to be *saturated with idempotents* if $\forall$ s $\in$ S, s^2 = s.

EXAMPLE V. 19: Consider $M_2(D_2)$, the ring of 2 x 2 diagonal matrices over the field Z_2. Taking involution to be the matrix transpose, then some short computations show that $M_2(D_2)$ is saturated with idempotents. In general, $M_2(D_2)$, the ring of n x n diagonal matrices over the field Z_2 is saturated with idempotents.

The next result is from Charles Lanski's paper, *"Rings With Involution Which Are I-rings"*:

THEOREM V. 20: Assume R is a ring with involution * which has no nilpotent elements and that for each s$\in$ S, the set of symmetric elements, there is a y $\in$ R such that sys = s. Then R is MP1.

COROLLARY V. 21: If R is a simple ring with no nonzero nilpotent elements and involution "*" and S $\subseteq$ Z(R) then R is MP1 and trivially MP4.

COROLLARY V. 22: If R is a ring with involution "*" having no nonzero nilpotent elements and all elements of S are invertible, then R is MP1.

COROLLARY V. 23: If R is a commutative, noetherian ring with involution "*", MP4, and S is saturated with idempotents, then R is MP1.

PROOF OF Corollary V. 21: Since R is simple, Z(R) is a field and thus for each s $\in$ S $\exists$ y such that sys = s. Hence, by Lanski's theorem, R is MP1. Also, R is trivially MP4 since for each nonzero a $\in$ R, aa* is nonzero.

PROOF OF Corollary V. 22: Apparent.

PROOF OF Corollary V. 23: R is an MP2 ring and since R is commutative, noetherian MP2, then R is a direct sum of fields. Hence, R is MP1.

OBSERVATION: If R is a commutative ring with involution * then S is multiplicatively closed when all of its elements are invertible. If R is also MP4 then R has no proper left ideals. All elements of R are left invertible; hence R is a field.

PROPOSITION V. 24: Let R be a ring with identity and involution *. If all the elements of s are invertible and all ideals I are such that $I^* = I$, then every homomorphic image of R is MP1.

PROOF: Straightforward. R/I is 2-torsion free and S has no nonzero nilpotent elements. By Lanski's result, R is MP1. Then with the given hypotheses, each homomorphic image is MP1 induced.

PROPOSITION V. 25: Let R be a ring with involution *; if R is prime and has no nonzero nilpotent elements, then R is MP4.

PROOF: By the hypothesis, R has no nonzero divisors of zero. Hence, given a $\in R$, aa^* is symmetric and R is MP4.

THEOREM V. 26: A direct sum of MP4 rings is MP4.

PROOF: Let $R_1, \ldots, R_n$ be MP4 rings with $*_i$ being the involution on R_i respectively. Let $R = \oplus \sum_{i=1}^{n} R_i$ and define the involution $*: R \to R$ in this manner: r

$\in R$ has the representation $r = (r_1, \ldots, r_n)$ where $r_i \in R_i$ and $r^* = (r_1^{*_1}, \ldots, r_n^{*_n})$. Let S_i be the set of symmetric elements of R_i. The set S of symmetric elements of R is represented then as $(S_1, \ldots, S_n)$. Now to show R is MP4, let $a \in R$, $a \neq 0$ so $a = (a_1, \ldots, a_n)$ where $a_i \in R_i$ and not all a_i are zero. Since each R_i is MP4, $\exists r_i \in R_i$, $r_i \neq 0$ such that $a_i r_i \in S_i$, not all $a_i r_i$ equal zero, and $(r_i a_i)^{*_i} = r_i a_i$. Now set r $= (r_1, \ldots, r_n)$. Then $ar = (a_1 r_1, \ldots, a_n r_n) = ((a_1 r_1)^{*_1}, \ldots, (a_n r_n)^{*_n}) = (ra)^* \in S$ and ar $\neq 0$. Hence, R is MP4.

THEOREM V. 27: A direct summand of an MP4 ring is MP4 under the induced involution.

PROOF: Let R be an MP4 ring with involution * and T a direct summand of R. Then $R = T \oplus V$ where $T \cap V = \{0\}$. If S is the set of symmetric elements of R then $S \cap T$ is the set of symmetric elements of T and $S \cap V$ is the set of symmetric elements of V. Assume $t \in T$, $t \neq 0$. Thereby $t' = (t,0) \in R$. Since R is MP4, $\exists r \in R$, $r \neq 0$ such that $t'r \in S$ and $t'r \neq 0$. Using the representation $r = (\rho, v)$ then $t\rho \in S \cap T$, $t\rho \neq 0$ since otherwise $t'r = 0$. So $(t\rho)^{*_T} = t\rho$ where $*_T$ is the involution on T induced by restricting the involution $*: R \rightarrow R$ to T. Hence, T is MP4.

VE. LOCALIZATION OF AN MP4 RING

THEOREM V. 28: Let R be a ring with identity and involution *. Further let S be the set of symmetric elements of R and set $S' = S - \{0\}$. Assume F is a saturated multiplicatively closed subset of R such that $S' \subseteq F$. If R is MP4, then R_F is MP4.

PROOF: Let a/b be an arbitrary nonzero element of R_F. Since $a,b \in R$ and R is MP4, there exists r, $p \in R$ such that $ar \in S$, $ar \neq 0$ and $bp \in S$, $bp \neq 0$. Se t $= r/p$. Then $(a/b)t = ar/bp = (ar)^*/(bp)^* \in S_F$, the set of symmetric elements of R_F. Hence, R_F is MP4.

THEOREM V. 29: Let R be a ring with involution "*", S its set of symmetric elements and T a ring with involution "$*$". Let $\varphi:R \rightarrow T$ be a ring homomorphism such that for each $r \in R$, $\varphi(r^*) = \varphi(r)^*$ and $S \cap \ker\varphi - \{0\}$. If R is MP4, then the image of R in T is MP4.

PROOF: Let $a' \in \varphi(R) \leq T$, $a' \neq 0$. Then $\exists\ a \in R$, $a \neq 0$, such that $\varphi(a) = a'$. But R MP4 $\Rightarrow \exists\ r \in R$ such that $ar \in S$, the symmetric elements of R and ar $\neq 0$. Then $\varphi(ar) = \varphi(a)\ \varphi(r) = \varphi((ar)^*) = \varphi(r^*a^*) = \varphi(r^*)\varphi(a^*) = \varphi(r)^*\ \varphi(a)^* = (\varphi(a)\varphi(r))^*$. Hence, $\varphi(a)\varphi(r) = a'\ \varphi(r) = (a'\varphi(r))^*$. Now $a'\varphi(r) \neq 0$ since otherwise $a'\varphi(r) = 0 \Rightarrow \varphi(r^*a^*) = \varphi(ar) = 0 \Rightarrow ar \in \ker\varphi \Rightarrow ar = 0$. Contradiction!! Therefore, $a'\varphi(r) \neq 0$ and $\varphi(R)$ is MP4 under the involution "$*$".

COROLLARY V. 30: If R is richly MP4 then any homomorphic image of R is MP4 with the induced involution.

EXAMPLE V. 31: Let R be the ring $\{ \begin{bmatrix} x & 0 & 0 \\ 0 & y & 0 \\ 0 & 0 & z \end{bmatrix} \mid x,y,z \in R$, the field of real numbers$\}$. Now set $A = \begin{bmatrix} a & 0 & 0 \\ 0 & b & 0 \\ 0 & 0 & c \end{bmatrix}$ and let the homomorphism φ be left multiplication by $\begin{bmatrix} 1 & 0 & 0 \\ 0 & 1 & 0 \\ 0 & 0 & 0 \end{bmatrix}$. If $A \neq 0$ then $\varphi(A) = \begin{bmatrix} a & 0 & 0 \\ 0 & b & 0 \\ 0 & 0 & 0 \end{bmatrix} \neq 0$. Let the involution * on R be matrix transpose. Then $A^* = A$ so that all elements of R are symmetric. Hence, R is MP4. In fact, R is richly MP4. To see this, let $B = \begin{bmatrix} b_1 & 0 & 0 \\ 0 & b_2 & 0 \\ 0 & 0 & b_3 \end{bmatrix}$; then $BB^* = \begin{bmatrix} b_1^2 & 0 & 0 \\ 0 & b_2^2 & 0 \\ 0 & 0 & b_3^2 \end{bmatrix} = 0$ if and only if $b_1 = b_2 = b_3 = 0$ if and only if $B = 0$. Now

$\ker\varphi \supseteq \{ \begin{bmatrix} 0 & 0 & 0 \\ 0 & 0 & 0 \\ 0 & 0 & g \end{bmatrix} \mid g \in R$, the field of real numbers$\}$ and ker $\varphi \cap R \neq 0$. In addition, $\varphi(R) = \{ \begin{bmatrix} u & 0 & 0 \\ 0 & v & 0 \\ 0 & 0 & 0 \end{bmatrix} \mid u,v \in R$, the field of real numbers$\}$. Suppose the involution $\star$ on $\varphi(R)$ is matrix transpose also. If $C \in R$, $C \neq 0$, then $\varphi(C^*) = \varphi(C^T) = \varphi(C) \star = \varphi(C)^T$. (*Note: Superscript "T" is the standard notation for matrix transpose*). Hence, under matrix tranpose, $\varphi(R)$ is MP4.

However, let $\star$, the involution on $\varphi(R)$ be interchanging the (1,1) and (2,2) entries of an element of $\varphi(R)$. Now if $D = \begin{bmatrix} 0 & 0 & 0 \\ 0 & d & 0 \\ 0 & 0 & 0 \end{bmatrix} \neq 0$ then $D\star = D$ if and only if $d = 0$ if and only if $D = 0$. Hence, $\varphi(R)$ under the diagonal-entry exchange involution $\star$ is not MP4.

THEOREM V. 32: Let R be an MP4 ring with identity and involution "*". Then

R[x] is MP4 with involution defined by: given $f \in R[x]$ where $f = \sum_{k=0}^{n} a_k x^k$

then $f^* = \sum_{k=0}^{n} (a_k)^* x^k$.

PROOF: Let $g \in R[x]$, $g \neq 0$ where without loss of generality, g has a nonzero constant term. Too see this, suppose $h \in R[x]$ has the form $a_m x^m + a_{m+1} x^{m+1} + \ldots + a_n x^n$. Then $h = x^m (a_m + a_{m+1} x + \ldots + a_n x^{n-m}) = (a_m + a_{m+1} x + \ldots + a_n x^{n-m}) x^m$. If R[x] is MP4 then $\exists p \in R[x]$ $(p \neq 0)$ such that hp is nonzero symmetric. Then $(hp)^* = p^* h^* = p^* (a_m + a_{m+1} x + \ldots + a_n x^{n-m})^* (x^m)^* = (a_m + a_{m+1} x + \ldots + a_n x^{n-m})^* (x^m)^* p^* = x^m (a_m + a_{m+1} x + \ldots + a_n x^{n-m})^* p^* = x^m p(a_m + a_{m+1} x + \ldots + a_n x^{n-m})$ since $(x^m)^* = 1^* x^m = x^m$ because $1 = 1^*$. Hence, g may be assumed to have a nonzero constant term. So let $g = a_0 + a_1 x^1 + \ldots + a_n x^n$ where $a_i \in R$ for all i. Then $g^* = (a_0)^* + (a_1)^* x^1 + \ldots + (a_n)^* x^n$.

Now $gg^* = a_0(a_0)^* + [a_0(a_1)^* + a_1(a_0)^*] x^1 + \ldots + [a_0 a_n^* + a_1 a_{n-1}^* + \ldots + a_{n-1} a_1^* + a_n a_0^*] x^n$.

Since at least $a_0(a_0)^* \neq 0$ and all the coefficient expressions of the positive powers of x are symmetric expressions, then gg^* is symmetric and nonzero. Consequently, R[x] is MP4.

COROLLARY V. 33: If R is an integral domain with involution "*", then R[x] is MP4.

PROOF: Since R is an integral domain, $aa^* = 0$ if and only if $a = 0$ which means that R is richly MP4. Now using the induced polynomial involution on R[x], then by the theorem above, R[x] is MP4.

VF. MATRIX RING OVER AN MP4 RING

THEOREM V. 34: If $M_n(R)$ is MP4 then R is MP4.

PROOF: Let $a \in R$, $a \neq 0$. Put $A = \begin{bmatrix} a & 0 & \ldots & 0 \\ 0 & 0 & \ldots & 0 \\ \cdot & & & \\ \cdot & & & \\ 0 & 0 & \ldots & 0 \end{bmatrix} \neq 0$ in $M_n(R)$. Then $\exists X \in M_n(R)$,

$X \neq 0$ such that $(AX)^* = AX \neq 0$. If $X = (x_{ij})$, then $AX =$
$$\begin{bmatrix} a & 0 & . & . & . & 0 \\ 0 & 0 & . & . & . & 0 \\ & . & & & & \\ & . & & & & \\ 0 & 0 & . & . & . & 0 \end{bmatrix}$$

$$\begin{bmatrix} x_{11} & . & . & . & x_{1n} \\ & . & & & \\ & . & & & \\ x_{n1} & . & . & . & x_{nn} \end{bmatrix} = \begin{bmatrix} ax_{11}\ ax_{12}\ .\ .\ .\ ax_{1n} \\ 0.\ .\ .\ .\ .\ .\ .\ .\ .\ 0 \\ . \\ . \\ 0.\ .\ .\ .\ .\ .\ .\ .\ 0 \end{bmatrix}.$$

Then $\begin{bmatrix} ax_{11}\ ax_{12}\ .\ .\ .\ ax_{1n} \\ 0.\ .\ .\ .\ .\ .\ .\ .\ 0 \\ . \\ . \\ 0.\ .\ .\ .\ .\ .\ .\ .\ 0 \end{bmatrix} = AX = (AX)^* = \begin{bmatrix} (ax_{11})^*\ 0\ .\ .\ .\ .\ .\ .\ 0 \\ (ax_{12})^*\ 0\ .\ .\ .\ .\ .\ .\ 0 \\ . \\ . \\ (ax_{1n})^*\ 0\ .\ .\ .\ .\ .\ .\ 0 \end{bmatrix}.$

Hence, $ax_{1j} = 0$, $\forall j > 1$. Since $AX \neq 0$, must have $ax_{11} \neq 0$. Thus, $ax_{11} = (ax_{11})^*$ and thereby R is MP4.

THEOREM V. 35: Assume R is a ring with identity and involution * and suppose R contains no nonzero nilpotent elements. Then R is MP4 if and only if $M_n(R)$ is MP4.

PROOF: Suppose R is MP4 and let $A \in M_n(R)$, $A \neq 0$. Let a_{ij} be the first nonzero entry of A. Without loss of generality, assume $j = 1$ since A is column equivalent to the matrix resulting from switching

the jth and 1st columns of A. Hence, $A = \begin{bmatrix} 0 \\ . \\ . \\ . \\ 0 \\ a_{i1} \\ a_{i+1,1} \\ . \\ . \\ a_{n1} \end{bmatrix}$ and set $E_1 =$

$$\begin{bmatrix} 1 & 0 & . & . & . & . & 0 \\ 0 & . & . & . & . & . & 0 \\ . & & & & & & \\ . & & & & & & \\ 0 & . & . & . & . & . & 0 \end{bmatrix} . \text{ Then } AE_1 = \begin{bmatrix} 0 & 0 & . & . & . & . & . & 0 \\ . & & & & & & \\ . & & & & & & \\ 0 & 0 & . & . & . & . & . & 0 \\ a_{i1} & 0 & . & . & . & . & 0 \\ a_{i+1,1} & 0 & . & . & . & 0 \\ . & & & & & & \\ . & & & & & & \\ a_{n1} & 0 & . & . & . & . & 0 \end{bmatrix} . \text{ Since } R \text{ is MP4, } \exists r_1, \ldots, r_n$$

in R

such that $a_{i1} r_j = s_j, \ldots, a_{in} r_n = s_n$ where $s_j, s_{j+1}, \ldots, s_n$ are all nonzero symmetric elements in S. Now

$$\text{set } P = \begin{bmatrix} r_i & 0 & . & . & . & . & 0 \\ 0 & . & . & . & . & . & 0 \\ . & & & & & & \\ . & & & & & & \\ 0 & . & . & . & . & . & 0 \end{bmatrix} . \text{ Then } AE_1 P = \begin{bmatrix} 0 & 0 & . & . & . & . & 0 \\ . & & & & & & \\ . & & & & & & \\ 0 & 0 & . & . & . & . & 0 \\ s_1 & 0 & . & . & . & . & 0 \\ r_i a_{i+1,1} & 0 & . & . & . & 0 \\ . & & & & & & \\ . & & & & & & \\ r_i a_{n1} & 0 & . & . & . & . & 0 \end{bmatrix} . \text{ Now set } C -$$

$AE_1 P$;

then CC^* has $s_i^2 \neq 0$ in the (i,i) position since R has no nonzero nilpotent elements and C^* is the involution transpose of C. Hence, set $X = E_1 PC^*$. Now $X \neq 0$ since otherwise $AX = CC^* = 0 \Rightarrow s_i^2 = 0$ which is a contradiction. Therefore, $M_n(R)$ is MP4.
Now assume $M_n(R)$ is MP4. Then by the previous theorem, R is MP4.

COROLLARY V. 36: Assume R is a ring with identity and involution * and suppose that each element of S, the set of symmetric elements, is idempotent. Then R is MP4 if and only if $M_n(R)$ is MP4.

PROOF: A nonzero idempotent element cannot be nilpotent.

DEFINITION V. 37: An associative ring R is called *formally real* if for any finite set of elements $\{a_i\}$ ($1 \leq i \leq n < \infty$) in R such that $a_1^2 + a_2^2 + \ldots + a_n^2 = 0$ then $a_1 = a_2 = \ldots = a_n = 0$.

THEOREM V. 38: Let R be an MP4 ring with involution "*"; also, let R be formally real and S saturated with idempotents. Then $M_n(R)$ is MP4 if $\forall\ a \in R$, $aa^* = 0$ implies $a = 0$.

PROOF: Suppose for a nonzero $A \in M_n(R)$, $AA^* = 0$. Then in particular, for $A = (a_{ij})$:

$$a_{11}a_{11}^* + a_{21}a_{21}^* + \ldots + a_{n1}a_{n1}^* = 0$$

.

.

.

$$a_{1n}a_{1n}^* + a_{2n}a_{2n}^* + \ldots + a_{nn}a_{nn}^* = 0$$

But $a_{ij}a_{ij}^*$ is symmetric; hence, $(a_{ij}a_{ij}^*)^2 = a_{ij}a_{ij}^*$.
Then, $(a_{11}a_{11}^*)^2 + \ldots + (a_{n1}a_{n1}^*)^2 = 0$

.

.

.

$$(a_{1n}a_{1n}^*)^2 + \ldots + (a_{nn}a_{nn}^*)^2 = 0$$

which implies $a_{11}a_{11}^* = \ldots = a_{nn}a_{nn}^* = 0$ since R is formally real. Thus, $\forall i,j\ a_{ij} = 0$ by hypothesis and $A = 0$. Contradiction! $A \neq 0$. Hence, $M_n(R)$ is MP4.

COROLLARY V. 39: If R is richly MP4 and formally real then $M_n(R)$ is richly MP4.

VG. MP4, MP2 AND MP1 EQUIVALENCE

THEOREM V. 40: Let R be a nontrivial ring with involution "*". Assume S, the set of symmetric elements, is nonzero and that each nonzero element of S is invertible. Further, assume for any nonzero idempotent f there exists an element p in R such that fp is a nonzero symmetric element. Then the following statements are equivalent:

(1) R is MP4.
(2) R is MP2
(3) R is MP1
(4) R is a division ring.

PROOF: (1) $\Rightarrow$ (4) Assume R is MP4 and a $\in$ R (a $\neq$ 0). Since R is MP4, there exists an r in R such that ar lies in S and ar $\neq$ 0. But ar symmetric implies that ar is invertible. Hence, there exists an y in R such that ary = 1. Set x = ry. Then x $\neq$ 0 since otherwise 0 = ax = 1. Contradiction! Hence, ax = 1 and R is a division ring Now, (4) $\Rightarrow$ (3) Since R is a division ring, R is MP1. (3) $\Rightarrow$ (2) R is MP1 and hence it is well-known to be MP2. (2) $\Rightarrow$ (1) Let a be an arbitrary element of R, a $\neq$ 0. Since R is MP2, there exists an r in R such that ar = e which is a nonzero idempotent element. By hypothesis there exists an p in R such that ep is nonzero symmetric. Set x = rp. Now x $\neq$ 0 since otherwise ax = ep = 0. Contradiction! Hence, ax is nonzero symmetric and R is MP4.

VH. INTEGRAL EXTENSION OF AN MP4 RING

THEOREM V. 41: Let R be an associative MP4 ring with unity and involution "*". Let Q be an integral extension of R. Then Q is MP4 under the involution "*".

PROOF: Since Q is an integral extension of R, for arbitrary q $\in$ Q, there exists a positive integer n and elements a_{n-1}, a_{n-2}, ..., a_0 in R such that $q^n + a_{n-1}q^{n-1} + \ldots + a_1q + a_0 = 0$. Then adding $-a_0$ to both sides of this equation yields the following:

$$-a_0 = q^n + a_{n-1}q^{n-1} + \ldots + a_1q$$

Since R is MP4, there exists an $r_0 \in$ R ($r_0 \neq 0$) such that $(-a_0) r_0 \neq 0$ and $(-a_0) r_0 \in$ S, the set of symmetric elements of R. Consequently,

$$(q^n + a_{n-1}q^{n-1} + \ldots + a_1q) r_0 \in S$$

Since q is common to each term of the expression in the second factor of this equation, by the distributive law

$$q(q^{n-1} + a_{n-1}q^{n-2} + \ldots + a_1) r_0 \in S$$

Now set $q' = (q^{n-1} + a_{n-1}q^{n-2} + \ldots + a_1) r_0$. Then $q' \neq 0$ (else $(-a_0) r_0 = 0$) and $qq' \in$ S. Hence, for arbitrary q $\in$ Q $\exists$ q' $\in$ Q such that qq' is symmetric. Then Q is MP4 under the involution "*" of R.

VI. MP4 MISCELLANEOUS RESULTS

DEFINITION V. 42: Call a nonzero ideal J MP4-induced if J is MP4 in the sense that given $a \in J$, $a \neq 0$, then $aJ \cap S \neq 0$ where S is the set of symmetric elements in R.

THEOREM V. 43: Assume R is a ring with involution "*" and identity. Suppose that for each nonzero ideal A, A^* is an MP4-induced commutative subring of R. Then R is MP4.

PROOF: Let a be an arbitrary nonzero element of R and $<a>$ denote the ideal generated by a. The ideal $<a>^*$ is a commutative MP4-induced subring of R. An arbitrary element of $<a>^*$ has the form $\sum_{i=1}^{n} r_i a^* s_i$. By hypothesis, since

$<a>^*$ is MP4-induced, there exists an t in $<a>^*$ such that $(\sum_{i=1}^{n} r_i a^* s_i)t \in S$ and

$(\sum_{i=1}^{n} r_i a^* s_i)t \neq 0$. Since R has an identity and $<a>^*$ is commutative, $(\sum_{i=1}^{n} r_i a^* s_i)t$

$= t(\sum_{i=1}^{n} r_i a^* s_i)$. But $s_i t \in <a>^*$ and hence commutes with a^* so that $(\sum_{i=1}^{n} r_i a^* s_i)t$

$= (\sum_{i=1}^{n} r_i s_i t)a^* = a^*(\sum_{i=1}^{n} r_i s_i)t = t(\sum_{i=1}^{n} r_i s_i t)a^*$ since $\sum_{i=1}^{n} r_i s_i t \in <a>^*$ and com-

mutes with a^*. Hence, $t(\sum_{i=1}^{n} r_i a^* s_i) = a^*(\sum_{i=1}^{n} r_i s_i)t$. But, $t(\sum_{i=1}^{n} r_i a^* s_i) \in S \Rightarrow$

$t(\sum_{i=1}^{n} r_i s_i)a^* \in S$. Then $(t(\sum_{i=1}^{n} r_i s_i)a^*)^* = a(\sum_{i=1}^{n} r_i s_i)^* t^* \in S$. Now set x =

$(\sum_{i=1}^{n} r_i s_i)^* t^*$. Hence, $x \neq 0$ since otherwise $t(\sum_{i=1}^{n} r_i s_i)a^* = 0$ contrary to deriva-

tion. Hence $ax \neq 0$ and $ax \in S$. Thus, R is MP4.

DEFINITION V. 44: Now let R be a ring with involution "*". Let G be a multiplicative group. If $x \in \Gamma_R(G)$, (the group ring of R over G), then define x by x $= \sum_g \mathbf{X_g g}$ where $X_g \in R$ and where the involution $*:\Gamma_R(G) \to \Gamma_R(G)$ is defined

by $x^* = \sum_g (\mathbf{X_g})^* \mathbf{g}^{-1}$ where $g \in G$.

PROPOSITION V. 45: If R is a MP4 domain and G is cyclic-finite, then $\Gamma_R(G)$ s MP4.

PROOF: Let $x = \sum_{i=0}^{n-1} \mathbf{X_i g^i}$ where the order of G, $o(G) = n$. Now set $y =$

$\sum_{i=0}^{n-1} (\mathbf{X_i})^* \mathbf{g}^{n-i}$. Then it is readily computed that xy is nonzero symmetric.

Hence, $\Gamma_R(G)$ is MP4.

PROPOSITION V. 46: Let R be a ring with involution *. If for each a in R, $\text{rad}(aR) \cap S \neq 0$, then R is MP4.

PROOF: If x lies in $\text{Rad}(aR) \cap S$, then $x^* = x$ and $x^n = y$ which lies in aR. Say $x^n = ar'$. Then $(x^n)^* = (ar')^* = (x^*)^n = x^n = ar'$.

VJ. MP3 AND MP4 IMPLICATIONS

THEOREM V. 47: Let R be an MP3 ring with involution * and assume for each nonzero a in R there exists an element c in R such that ca* is a nonzero central element; then R is MP4.

PROOF: Let a be an arbitrary nonzero element of R. Then by hypothesis, $\exists c \in R$ such that $ca^* \in Z(R)$, the center of R. But R MP3 implies $\exists r \in R$ such that $rca^* \in S$, the set of symmetric elements of R and $rca^* \neq 0$. Since rca^* is symmetric, $(rca^*)^* = ac^*r^* \in S$. Now set $x = c^*r^*$. Note $x \neq 0$ since otherwise $x^* = 0$ and $x^*a^* = rca^* = 0$. Contradiction! Hence, $ax \neq 0$ and $ax \in S$. Therefore, r is MP4.

THEOREM V. 48: Let R be an MP4 ring with involution * and assume for each nonzero a in R there exists an element c in R such that a*c is a nonzero central element; then R is MP3.

PROOF: Let a be an arbitrary nonzero element of R. Then by hypothesis, $\exists c \in$ R such that $a*c \in Z(R)$. But R MP4 implies that $\exists r \in R$ such that $a*cr \in S$, the set of symmetric elements in R and $a*cr \neq 0$. Since $a*cr$ is symmetric, $(a*cr)*$ $= r*c*a \in S$. Now set $x = r*c*$. Note $x \neq 0$ since otherwise $x* = 0$ and $a*x* =$ $a*cr = 0$. Contradiction! Hence, $xa \neq 0$ and $xa \in S$. Therefore, R is MP3.

COROLLARY V. 49: Let R be a ring with involution * and suppose that all symmetric elements are central. Then R is richly MP4 if and only if R is richly MP3.

PROOF: Assume R is richly MP4 and let $a \in R$ be arbitrary nonzero. Since R is MP4 and since there exists an c in R such that $a*c$ lies in $Z(R)$, the center of R, namely $(a*)* = a$, then by the previous theorem, R is MP3. But $a*a \in S$ and $a*a \neq 0$ if and only if $a \neq 0$. Hence, R is richly MP3. Now assume R is richly MP3. Let a be an arbitrary nonzero element of R. Then $a* \neq 0$ and $(a*)*a* =$ $aa* \in Z(R)$ since $aa*$ is symmetric. By the previous theorem, R is MP4 since R is MP3 and there exists an c in R such that $ca*$ is nonzero central. But $aa* \neq$ 0 if and only if $a \neq 0$ and R is richly MP4.

EXAMPLE V. 50: A commutative ring R with involution * such that the set of symmetric elements $S \subseteq Z(R)$, the center of R.

Let $R' = F(x, y)$ whre F is a field and x, y are noncommuting variables. Further, let the involution * on $F(x, y)$ be switching the x and y variables, e.g., $(x^2 +$ $y)* = y^2 + x$. Define $R = F(x,y) / (x^i y^i, y^i x^i, x^m, y^m)$ where $i \geq 1$, $m \geq 2$ and $(x^i y^i,$ $x^i y^i, x^m, y^m)$ is the ideal generated by all specified powers of xy, yx and x and y. Also, let the involution * on R be the induced involution from R'. Then a typical element of R has the form $r = a + bx + cy$ where a, b, $c \in F$. In particular, S, the set of symmetric elements of R, is $\{\alpha, \beta(x + y), \gamma + \delta(x + y) | \alpha, \beta, \gamma, \delta \in F\}$. Demonstrably, $\alpha* = \alpha$, $\beta(x + y)* = (x* + y*)\beta* = (y + x) \beta = \beta(x + y)$ and $(\gamma$ $+ \delta(x + y))* = \gamma* + (\delta(x + y))* = \gamma* + (x* + y*)\delta* = \gamma + (y + x)\delta = \gamma + \delta(x +$ y). Next, the claim is that $S \subseteq Z(R)$. To see this, first note that $\alpha \in F$ commutes with everything, that is to say, $\alpha r = r\alpha$, $\forall r \in R$. Now $\beta(x + y)r = \beta(x + y)(a +$ $bx + cy) = \beta a(x + y) + \beta bx^2 + \beta byx + \beta cxy + \beta cy^2 = \beta a(x + y) = a\beta(x + y) =$ $r\beta(x + y)$, $\forall r \in R$. Finally, $(\gamma + \delta(x + y))r = (\gamma + \delta(x + y))(a + bx + y) = \gamma(a + bx$

$+cy) + \delta a(x + y) + \delta bx^2 + \delta cxy + \delta byx + \delta cy^2 = \gamma(a + bx + cy) + \delta a(x + y) = (a + bx + cy)\, \gamma + a\delta(x + y) = r(\gamma + \delta(x + y))$. Hence, $S \subseteq Z(R)$.

THEOREM V. 51: Let R be a commutative ring. Then R is MP4 if and only if R is MP3.

PROOF: Assume R is MP4. Let a be an arbitrary nonzero element of R. Since R is MP4 there exists an r in R, $r \neq 0$ such that $ar = (ar)^* = r^*a^* \in S$, the set of symmetric elements. Since R is commutative, $r^*a^* = a^*r^* = (ra)^* = ra$ lies in S. Hence, R is MP3. Now assume R is MP3. Let a be an arbitrary nonzero element of R. Since R is MP3, there exists an r in R, $r \neq 0$ such that $ra = (ra)^* = a^*r^*$ lies in S, the set of symmetric elements. Since R is commutative, $a^*r^* = r^*a^* = (ar)^* = ar$ lies in S. Hence, R is MP4.

Before concluding this introductory treatise on Moore-Penrose rings, the author notes there remain some speculative questions regarding MP3 and MP4 rings. One open question is whether or not an arbitrary matrix ring over an MP3 (MP4) ring is MP3 (MP4). Moreover, further investigation of complete equicharacteristic local rings of dimension greater than one may prove to be a hidden treasure of new mathematical wisdom.

VI. CONCLUSION

Overall, many interesting properties have been surveyed for the ring analogue of the Moore-Penrose matrix equations: Given a in R, a nonzero, there exists an x in R, x nonzero such that

$$(1) \ axa = a \qquad \text{MP1 ring}$$
$$(2) \ xax = x \qquad \text{MP2 ring}$$
$$(3) \ (xa)^* = xa \qquad \text{MP3 ring}$$
$$(4) \ (ax)^* = ax \qquad \text{MP4 ring}$$

where "*" denotes a ring involution. The MP1 and MP2 rings are worthy of investigation of their algebraic properties not only because in a matrix ring setting, such matrices can be applied to solving linear statistical models equations sometimes not of full rank to obtain uniform minimum-variance unbiased estimators, but also because MP2 rings envelope the MP1 rings with subtle or sheer distinctions that vanish under finite order and chain conditions. Both MP1 and MP2 rings abound with idempotents and the relation of these idempotents to their ring ideal structure lead to properties which characterize them dramatically. Similarly, MP3 and MP4 rings are defined in terms of their one-sided ideals intersecting nontrivially with the set of symmetric elements, usually denoted S, determined by the involution on such rings. To a large extent, the size of S and algebraic properties imposed on it characterize a great deal about the structure of an MP3 or MP4 ring. Since MP3 and MP4 rings differ in definition by a lettering transposition, it is not surprising that many of the algebraic results for MP3 rings hold true for MP4 rings. Naturally, in the commutative ring case, both MP3 and MP4 are equivalent. Whereas this Volume I book has attempted to thoroughly examine the structure of Moore-Penrose rings, their computational clarity in the theory of physical science, engineering applications, business and economic modeling, and countless scientific venues are their appeal by applied scientists and engineers. Hence, there is a need to introduce the novice reader in the applied mathematical sciences to an enriched approach that would render hopefully a greater appreciation for these Moore-Penrose rings and their validated usefulness.

BIBLIOGRAPHY

1. Amitsur, S.A., "Rings With Involution", <u>Israel J. Math</u>, 6, 99–106 (1968).

2. Armendariz, E. P., and Fisher, Joe W., "Regular P. I. Rings", <u>Proc. Amer. Math. Soc.</u>, 39(1973), 247–251.

3. Cohen, I. S., "On The Structure and Ideal Theory of Complete Local Rings", <u>Trans. Amer. Math. Society</u>, Vol. 59 (1946), 54–55.

4. Dyer-Bennet, J., "On Finite Regular Rings", <u>Bulletin Amer. Math. Society</u>, 47(1941), 784–787.

5. Ehrlich, Gertrude, "Unit-Regular Rings", <u>Portugaliae Mathematica</u>, Vol. 27 (1968), 209–212.

6. Erickson, D. B., "Orders For Finite Noncommutative Rings", <u>Mathematical Notes</u>, April 1966, 376–377.

7. Fisher, Joe W. and Snider, Robert L., "Prime Von Neumann Regular Rings and Primitive Group Algebras", <u>Proc. Amer. Math. Society</u>, 44(1974), 244–249.

8. Fisher, Joe W. and Snider, Robert L., "On The Von Neuman Regularity Of Rings With Regular Factor Rings", <u>Pacific Journal of Mathematics</u>, Vol. 54, No. 1, 1974, 135–144.

9. Geddes, A., "A Short Proof of the Existence of Coefficient Fields for Complete Equicharacteristic Local Rings", <u>J. London Math. Society</u>, Vol. 29(1954), 334–341.

10. Gilmer, Robert, "On Polynomial And Power Series Rings Over A Commutative Ring", <u>Rocky Mountain J. Math.</u>, 5(1975), 151–175.

11. Goldie, A.W., "Non-commutative Principal Ideal Rings", <u>Arch. Math.</u>, 13, 213–221 (1962)

12. Goldie, A.W., "Localization In Non-commutative Noetherian Rings", <u>J. Algebra</u>, 5(1967), 89–105.

13. Goldie, A.W., Small, L.W., "A Note On Rings Of Endomorphisms", <u>J. Algebra</u>, 24(1973), 392–395.

14. Goodearl, K.R., "Some Representation Theorems For Involution Rings", <u>J. Algebra</u>, 14(1970), 299–311.

15. Goodearl, K.R., <u>Von Neumann Regular Rings</u>, Putnam Publishing Ltd., London, 1973.

16. Hartwig, Robert E., "An Application of Moore Penrose Inverse To Antisymmetric Relations", <u>Proc. Amer. Math. Society</u>, 78(1980), 181–186.

17. Hartwig, Robert E. and Luh, Jiang, "On Finite Regular Rings", <u>Pacific J. Math.</u>, 69(1977), 73–95.

18. Henriksen, Melvin, "On A Class Of Regular Rings That Are Elementary Divisor Rings", <u>Arch. Math (Basel)</u>, 24(1973), 133–141.

19. Herstein, I. N., <u>Noncommutative Rings</u>, The Carus Math. Monographs, Math. Association of America, 1968.

20. Jacobson, N., "Some Remarks On One-Sided Inverses", <u>Proc. Amer. Math. Society</u>, 1(1950), 352–355.

21. Kaplansky, Irving, <u>Fields and Rings</u>, Univ. of Chicago Press, Chicago, 1972.

22. Kaplansky, I., "Problems In The Theory Of Rings", <u>Amer. Math. Monthly</u>, 77 (1970), 445–454.

23. Lambek, Joachim, <u>Lectures On Rings And Modules</u>, Blaisdell Publishing Co., Waltham, Mass., 1966.

24. Lanski, Charles, "Chain Conditions In Rings With Involution", <u>J. London Math. Society</u>, 9(1974), 93–102.

25. Lanski, Charles, "On The Relationship Of A Ring And The Subring Generated By Its Symmetric Elements", <u>Pacific Jour. Math.</u> 44(1973), 581–592.

26. Lanski, Charles, "Rings With Involution Whose Symmetric Elements Are Regular", <u>Proc. AMS</u>, 33(1972), 264–270.

27. Lissner, D., "Matrices Over Polynomial Rings", <u>Trans. Amer. Math. Soc.</u>, 98 (1961), 285–305.

28. McCoy, N. H., "Subdirect Sum of Rings", <u>Bulletin Amer. Math. Society</u>, Vol. 53(1947), 856–877.

29. McCoy, N. H., "Subdirectly Irreducible Commutative Rings", Duke Math. <u>Journal</u>, Vol. 12(1945), 381–387.

30. McCoy, N., <u>Theory Of Rings</u>, Chelsea Publishing Co., Bronx, N.Y., 1973.

31. McDonald, B. R., "Diagonal Equivalence of Matrices Over a Finite Local Ring", <u>Journal of Combinatorial Theory</u>, 13(1972), 100–104.

32. Michler, G.O., "Idempotent Ideals In Perfect Rings", <u>Canad. J. Math.</u>, 21, 301–309 (1969)

33. Moore, E. H., "On The Reciprocal Of The General Algebraic Matrix", <u>Bull. Am. Math. Soc.</u>, 26, 394–395.

34. Nagata, Masayushi, <u>Local Rings</u>, Interscience Publishers, Inc., 1962, xi–xiii.

35. Narayan, Shanti, <u>A Textbook Of Matrices</u>, S. Chand & Co., New Delhi, 1964.

36. Penrose, R., "A Generalized Inverse For Matrices", <u>Proc. Camb. Philos. Soc.</u>, 51, 406–413.

37. Reineke, Joachim, "Commutative Rings In Which Every Proper Ideal Is Maximal", <u>Fund. Math.</u>, 97(1977), 229–231.

38. Saciada, E., Cohn, P.M., "An Example Of A Simple Radical Ring", <u>J. Algebra</u>, 5 (1967), 373–377.

39. Satyanarayana, M., "Subfields Of Matrix Rings", <u>Arch. Math. (Basel)</u>, 24(1979), 569–571.

40. Shanny, Robert F., "Regular Endomorphism Rings Of Free Modules", <u>J. London Math. Society</u>, Vol. 2(1971), 353–354.

41. Slover, R., "The Radical Of Row Finite Matrices", <u>J. Algebra</u>, 12(1969), 345–359.

42. Von Neumann, John, "On Regular Rings", <u>Proc. Nat. Academy Sciences</u>, 22(1936), 707–713.

43. Wong, Edward T., "Regular Rings And Integral Extension Of A Regular Ring", <u>AMS</u>, Vol. 33, 2, June 1972.

44. Zariski and Samuel, <u>Commutative Algebra</u>, Vol. 1, D. Van Nostrand Co., Inc., Princeton, N.J., 1958.

45. Zelinsky, D., "Raising Idempotents", <u>Duke Math. J.</u>, 21(1954), 315–322.

APPENDIX

The following is a generalization of the LEMMA (page 7) followed by a general theorem of a matrix ring over an MP2 ring.

THEOREM: Let R be a ring in which every nonzero principal left ideal of R contains a nonzero idempotent, then R is MP2.

If R contains a unity (or identity), then R is MP2 by Battle's Theorem. Now assume R does not contain a unity. Suppose $a \in R$, $a \neq 0$. According to Neal H. McCoy in his book, The Theory of Rings, the principal left ideal generated by a is given by the set $\{na + sa \mid n \in Z$, the ring of integers; $s, t \in R\}$ which he denotes by $<a>_L$. McCoy notes that $Ra = \{ ra \mid r \in R\}$ is itself a principal left ideal contained in $<a>_L$.

Now assume $Ra = 0$, the zero ideal. By hypothesis, $<a>_L$ contains a nonzero idempotent e where $e = na + sa$ for some $n \in Z$ and $s \in R$. Since e is idempotent, e^2 must equal $e \neq 0$. But $e^2 = e = (na + sa)(na + sa) = n^2a^2 + nasa + sana + sasa = 0$, since $Ra = 0$ implies that $a^2 = 0$ and $sa = 0$. This contradicts the fact that e is a nonzero idempotent. Thus, $Ra \neq 0$. Furthermore, Ra is a nonzero principal left ideal and must contain a nonzero idempotent, say f such that $f^2 = f$. Then $f \in Ra \subseteq <a>_L$. Now $f = ta$ for some $t \in R$.

THEOREM: Let R be a ring with identity. Then $M_n(R)$ is MP2 provided that R is MP2.

PROOF: Assume R is MP2. Let $A \in M_n(R)$ be nonzero. Suppose a_{kp} (k, $p \leq n$) is the first nonzero entry of A. Since R is MP2, there is an element $r \in R$ such that $ra_{kp} = e$ an idempotent. Let E be the matrix with an r in the (k,k) position and zeros elsewhere.

Then EA =
$$
\begin{bmatrix}
0 \ldots \ldots \ldots \ldots \ldots \ldots \ldots & 0 \\
\cdot & \cdot \\
\cdot & \cdot \\
0 \ldots \ldots e\, \mathbf{n}_{p+1} \ldots \mathbf{n}_k & \\
0 \ldots \ldots \ldots \ldots \ldots \ldots 0 & \\
\cdot & \cdot \\
\cdot & \cdot \\
0 \ldots \ldots \ldots \ldots \ldots \ldots 0 &
\end{bmatrix}
$$

Certainly, EA is in the principal ideal of $M_n(R)$ generated by A. Now postmul-

tiply the matrix EA by the matrix F which has e in the (p, p) position and zeros

elsewhere. Then EAF =
$$
\begin{bmatrix}
0 \ldots \ldots \ldots \ldots \ldots \ldots \ldots 0 \\
\cdot \qquad\qquad \cdot \\
\cdot \qquad\qquad \cdot \\
0 \ldots \ldots e\, 0 \ldots \ldots \ldots \ldots 0 \\
0 \ldots \ldots \ldots \ldots \ldots \ldots 0 \\
\cdot \qquad\qquad \cdot \\
\cdot \qquad\qquad \cdot \\
0 \ldots \ldots \ldots \ldots \ldots \ldots 0
\end{bmatrix}
$$

Now premultiply EAF by the matrix E_{pk} which has an e in both the (p,p)-th
and (k,k)-th position. Its function shall be to switch the kth and pth rows. Then
$E_{pk}EAF$ is the matrix whose (p, p)-th entry is e and zeros elsewhere. Hence
$E_{pk}EAF = I$ a matrix with e in the (p, p)-th position.
Note that $I^2 = I$ since e is idempotent. Now premultiply $E_{pk}EAF = I$ by F and
postmultiply $E_{pk}EAF = I$ by $E_{pk}E$ so that $FE_{pk}EAFE_{pk}E = FIE_{pk}E$. Evenmore,
$FE_{pk}EAFE_{pk}E = FIE_{pk}EAFIE_{pk}E$ since I is idempotent. Set $J = FIE_{pk}E$ so that JAJ
$= J$ and $M_n(R)$ is MP2. Note that $J \neq 0$ since otherwise $E_{pk}EAJAF = 0$ would
force e = 0 as a contradiction.

978-0-595-37806-7
0-595-37806-4

www.ingramcontent.com/pod-product-compliance
Lightning Source LLC
Chambersburg PA
CBHW021024180526
45163CB00005B/2098